# SUCCESS AT

# AQA

## PHYSICS B A2

Ken Price & Gerard Kelly

**OXFORD**

UNIVERSITY PRESS

# OXFORD
## UNIVERSITY PRESS

Great Clarendon Street, Oxford OX2 6DP

Oxford University Press is a department of the University of Oxford.
It furthers the University's objective of excellence in research,
scholarship, and education by publishing worldwide in

Oxford  New York

Auckland  Bangkok  Buenos Aires  Cape Town  Chennai
Dar es Salaam  Delhi  Hong Kong  Istanbul  Karachi  Kolkata
Kuala Lumpur  Madrid  Melbourne  Mexico City  Mumbai
Nairobi  São Paulo  Shanghai  Taipei  Tokyo  Toronto

Oxford is a registered trade mark of Oxford University Press
in the UK and in certain other countries

© Ken Price and Gerard Kelly 2001

British Library Cataloguing in Publication Data

Data available

ISBN 0 19 914802 3

10 9 8 7 6 5 4 3

Designed and typeset by Cambridge Publishing Management Ltd

Printed and bound by G. Canale & Co., Italy

The publisher would like to thank the Lawrence Berkeley national laboratory,
of the University of California, for their kind permission to reproduce the
photograph on page 125.

# Contents

# Introduction

## About This Book

In this book you will find:
- information about the examination papers you will take
- advice on how to tackle question papers effectively
- the definitions and facts that you must learn
- the principles that you should know, understand and be able to use in problems
- examples of the relevance of the principles in everyday life
- examples for you to try to test your understanding, that are included within topics and at the end of sections
- answers to the questions
- an index that enables you to locate information that you need more quickly.

The book covers all three **A2** modules. The specific practical skills in **Module 6** are covered in the work of the other two modules so this is presented first. The topics in **Modules 4** and **5** are clearly identified. Within the modules, some topics are in a different order from that in the specification. This has been done so that the progression from one topic to another is more logical.

Use this book as the framework for your studies, but find time to explore the ideas further.

You should use a variety of resources such as other textbooks, CD-ROMs and the internet. You will often find that one or other of the resources explains an idea in a way that enables you to understand the concepts more easily.

## Relationship of A2 to AS

The A2 course contains many new ideas. These are summarised in the contents list on the earlier pages.

As well as using this book to study the detail in the A2 course, you are advised to keep looking back and reviewing the material in the AS course. There are two reasons for this.
- During your A2 studies you will need to recall and apply some principles that you studied during the AS course.

- In Units 5 and 6 you may be asked questions on any of the topics in the AS and A2 parts of the course.

## Practical Work in A2

The basic skills of planning, implementing, analysing and evaluation should be developed further in this course.

These skills are developed through your practical assignments. Your practical work will also:
- improve your understanding of the principles in particular topics
- continue to develop your organisational and problem-solving skills.

Continue therefore to make good use of your practical time. Use every opportunity to discuss with others what you have found. Remember that clear and accurate communication of practical and theoretical ideas is vital for your future success in science.

## Key Skills

During A2 you may still be building up your Key Skills portfolio.

There are many opportunities through practical and research activities to gather the evidence you need.

You should be able to gather further evidence for your communication, numeracy and IT Key Skills portfolio.

You could also gather evidence for the three other Key Skills areas:
- **working with others** carrying out research or practical investigation
- **improving your own performance** by demonstrating that you have learned more about the topics by studying them in more detail than you have under guidance in the classroom
- **problem solving** when analysing new practical or theoretical situations.

## The A2 Examination

The A2 examination consists of three units.

**Table I.1**

|  | Unit 4 | Unit 5 | Unit 6 |
|---|---|---|---|
| Content | Module 4 | Module 5 + Modules 1, 2, & 4 | Module 6 Practical skills + Modules 1, 2, 4 & 5 |
| Time allowed | $1\frac{1}{2}$ h | 2 h | 3 h |
| Marks in paper | 75 | 100 | 78 |
| % of AL | 15 | 20 | 15 |

## Unit 4

This consists of short-answer questions and structured questions based on the content of **Module 4**. The unit is in the form of a question/answer booklet that you will have met in **Units 1** and **2**.

## Unit 5

This will test your understanding of the material in **Module 5**. This unit is also in the form of a question/answer booklet. The unit consists of structured questions and a comprehension question.

Some questions will be synoptic. This means that the marks are awarded in questions that test your ability to use and make connections between topics that have been covered in any of the modules in the specification, including those studied in AS. Approximately 75 of the 100 marks are 'synoptic marks'.

## Unit 6

This is a practical examination. It will test your ability to plan, implement, analyse and evaluate experiments.

The examination consists of three compulsory exercises that are discussed in more detail on page 15. You will answer the questions in an answer booklet.

This unit is partly synoptic, approximately one third of the marks being 'synoptic marks'.

## Examination hints

The following are a reminder of some points to note when taking an examination.

- Read the question carefully. Make sure you understand exactly what is required. Use the 'key words' that follow to guide you.
- If you find that you are unable to do a part of a question do not give up. The next part may be easier and may provide a clue to what you might have done in the part you found difficult.
- Note the number of marks per question as a guide to the depth of response.
- Underline or note the key words that tell you what is required (see page 8).
- Underline or note data as you read the question.
- Structure your answers carefully.
- Show all steps in calculations. Include equations you use and show the substitution of data. Remember to work in SI units.
- Make sure your answers are to suitable significant figures (usually 2 or 3).
- Include a unit unless the answer is a pure number or a ratio.
- Consider whether the magnitude of a numerical answer is reasonable for the context. If it is not, check your working.
- Draw diagrams and graphs carefully.
- Read data from graphs carefully; note scales and prefixes on axes.
- Keep your eye on the clock, but do not panic.
- If you have time at the end, use it. Check that your descriptions and explanations make sense. Consider whether there is anything you could add to an explanation or description. Repeat calculations to ensure that you have not made any mistakes.

# Revising for Examinations

There is no one method of revising that works for everyone. It is therefore important to find the approach that suits you best. The following rules may serve as general guidelines.

**Give yourself plenty of time:** Do not leave everything until the last minute. This reduces your chances of success and you could start to panic, which will reduce your concentration. Few people can revise everything the night before and still do well in an examination the next day.

**Plan your revision timetable:** Plan your revision timetable some weeks before the examination and follow it. Do not be side-tracked. Allow time for unforeseen problems that could arise, such as illness.

**Relax:** Concentrated revision is very hard work. It is as important to give yourself time to relax as it is to work. Build some leisure time into your revision timetable.

**Give yourself a break:** Work for about an hour then take a break for 15 to 20 minutes. Then go back to another productive revision period.

**Find a quiet corner:** Find the conditions in which you can revise most efficiently. Many people think they can revise in a noisy, busy atmosphere, but most cannot. Any distraction lowers concentration. Revising in front of a television does not generally work.

**Keep track:** Use a checklist and the specification to keep track of your progress. Mark off topics you have revised and feel confident with. Concentrate your revision on things you are less happy with.

**Make your own notes and use colours:** Revision is often more effective when you do something active rather than simply reading. The key ideas are in this book. Reading other books with different approaches might help you understand better.

**Concentrate on understanding:** Memorising definitions, facts, etc. is important, but try to understand how the ideas are applied.

**Practise answering questions:** As you finish each topic, try answering the questions within and at the end of each section, or try questions from past papers. Even when you have done them before, the act of remembering how they are done will help you learn.

# Key Words in Examination Papers

How you respond to a question can be helped by studying the following, which are the more common key words used in examination questions.

### Name

The answer is usually a technical term or its equivalent, consisting of one or two words.

### State

This requires a statement of a fact, a procedure, a principle or an equation without elaboration. Where further comment is needed the question will ask you to 'State and explain'.

### Explain

The important thing to note here is that a reason or explanation must be given, not just a description.

### List

You need to write down a number of points (each may be a single word) with no elaboration.

### Define/What is meant by?

'Define' requires you to give a precise meaning of a particular term. 'What is meant by' is used to emphasise that a formal definition is not required.

### Outline

You need to give a brief summary of the main points. The mark allocation is a good guide to the detail required.

### Describe

The answer is a description of an effect, experiment or possibly a graph shape. No explanations are required.

### Describe how you would

This is usually used in the context of experiment design. It requires you to say how an experiment could be done by you as a student working in your A2 laboratory.

### Suggest

This is used when it is not possible to give the answer directly from the facts that form part of the subject material in the specification. The answer may be based more on general understanding than on recall of learned material.

### Calculate/Determine a value for/Determine the magnitude of/Find

A numerical answer is required, usually using data given in the question. Remember to give your answer to a suitable number of significant figures and to give a unit.

### Justify

This is similar to 'Explain'. You will have made a statement and now have to provide a reason for giving that statement.

### Draw/Make a drawing/Sketch a diagram

You simply draw a diagram. You will usually be asked to label it. It is sensible to provide labelling even when this is not asked for.

### Sketch a graph

You need to draw the general shape of the graph on labelled axes. You should include enough quantitative detail to show relevant intercepts and/or whether the graph is exponential or some inverse function, for example.

### Plot

The answer will be an accurate plot of a graph on graph paper. Often there is also a question asking you to determine some quantity from the graph or to explain the shape of the graph.

### Estimate

You may need to use your knowledge and/or your experience to deduce the magnitude of some quantities to arrive at the order of magnitude for some other quantity defined in the question.

### Discuss

This will require an extended response in which you demonstrate your knowledge and understanding of a given topic.

### Show that

You will have been given either a set of data and a final value (which may be approximate) or an algebraic equation. You need to show clearly all the basic equations that you use and all the steps that lead to the final answer. Remember to quote the result of your working and then relate this to the value given in the question.

# Formulae

## Formulae

In your revision remember to:
- learn the formulae that are not on your formula sheet (**Note**: You have to learn more formulae for A2 than for AS)
- make sure that you know what the symbols represent in formulae that you have memorised and those on the formula sheet.

**You will be provided with the AS and A2 formula sheets. These are shown on Pages 11 and 12.**
Note that there are some amendments to the AS sheet since publication of 'Success at AQA Physics B AS'.

## Formulae to Remember

You will need to be able to recall and use the following mechanics formulae.

### New mechanics formulae in A2

Momentum $p = mv$
For motion in a circular path of radius $r$,

force toward centre $F = \dfrac{mv^2}{r}$

### Formulae from AS

$$v = \frac{s}{t} \qquad F = ma \qquad \rho = \frac{m}{V}$$

$$W = Fs \qquad P = \frac{W}{t} \qquad weight = mg$$

kinetic energy $= \frac{1}{2}mv^2$  $\Delta$(potential energy) $= mg\Delta h$

Table I.2 shows what each symbol represents.

**Table I.2**

| Symbol | Quantity | Symbol | Quantity |
|--------|----------|--------|----------|
| $s$ | Displacement | $P$ | Power |
| $v$ | Speed or magnitude of velocity | $W$ | Work done or energy transferred |
| $t$ | Time | $g$ | Acceleration of free fall |
| $F$ | Force | $h$ | Height |
| $m$ | Mass | $\Delta$ | Change in (e.g. $\Delta h$ = change in height) |
| $a$ | Acceleration | | |
| $\rho$ | Density | | |
| $V$ | Volume | | |

You will need to be able to recall and use the following electricity formulae:

### New electricity formulae in A2

$$C = \frac{Q}{V}, \qquad capacitance = \frac{charge}{potential\ difference}$$

For a transformer:

$$\frac{V_s}{V_p} = \frac{N_s}{N_p},$$

$$\frac{voltage\ across\ secondary\ coil}{voltage\ across\ primary\ coil} = \frac{number\ of\ turns\ on\ secondary}{number\ of\ turns\ on\ primary}$$

### Formulae from AS

$$\Delta Q = I\ \Delta t \qquad V = IR \qquad P = VI$$

$$p.d. = \frac{W}{Q} \qquad R = \frac{\rho l}{A} \qquad Energy = VIt$$

Table I.3 shows what each symbol represents.

**Table I.3**

| Symbol | Quantity | Symbol | Quantity |
|--------|----------|--------|----------|
| $Q$ | Charge | $W$ | Work done or energy transferred |
| $I$ | Current | | |
| $V$ | Potential difference or voltage | $\rho$ | Resistivity |
| | | $l$ | Length |
| $R$ | Resistance | $A$ | Cross-sectional area |
| $P$ | Power | $t$ | Time |

### Formula from AS

You will need to be able to recall and use the following wave formula:

$$v = f\lambda$$

where $v$ is velocity,
$f$ is the frequency
$\lambda$ is the wavelength.

### New fields formulae in A2

$$F = \frac{kQ_1Q_2}{r^2}$$

$$Electric\ force = \frac{constant \times charge1 \times charge2}{distance\ between\ point\ charges}$$

$$F = \frac{Gm_1m_2}{r^2}$$

$$gravitational\ force = \frac{gravitational\ constant \times mass1 \times mass2}{distance\ between\ point\ masses}$$

# Units

Physical quantities consist of a number and a unit.

The units go with the numbers and are treated as symbols in an equation to give a correct unit. They should also be transferred with the numerical values in graphical work.

## SI units

This is the system of units used in scientific work. **SI** stands for le **S**ystème **I**nternational. This system defines a number of base units.

## Base SI units

| | | |
|---|---|---|
| Mass | kg | kilogram |
| Length | m | metre |
| Time | s | second |
| Temperature | K | kelvin |
| Current | A | ampère |
| Amount of substance | mol | mole |
| Luminous intensity | cd | candela |

You will not need the candela in your advanced level studies.

## Derived units

Other units may be written in terms of these. It is not wrong to give the unit for a quantity in terms of the base units. The derived unit in terms of base units can be obtained if you know a formula for the quantity and the units of the other quantities.

EXAMPLE ONE: Force = mass × acceleration

$$F = ma$$

The unit of mass is kilogram, kg, and the unit for acceleration is $m\,s^{-2}$.

**A unit for force in base units is kg m$\,$s$^{-2}$**

Work done = force × displacement

$$W = Fs$$

The unit of $F$ is newton, N, and the unit of displacement $s$ is metre, m. A unit of work done or energy transferred is therefore N m.

**In terms of base units the unit for work done is kg m s$^{-2}$ m or kg m$^2$ s$^{-2}$.**

### Consider:

The usual unit for permittivity is farad per metre. In terms of base units this would be $A^2\,s^4\,kg^{-1}m^{-2}$! As seen in this example, the unit in terms of base units becomes cumbersome and it is easy to make mistakes. You should learn the appropriate SI unit and use it where relevant.

However, if you are really in doubt about the correct or usual SI unit, it is better to work it out in terms of base units than to give no unit at all.

## Checking equations

If an equation is correct, then the unit of each term, on both sides of the equation, must be the same.

EXAMPLE TWO: Could the following equation be a correct equation?

$$E = mgh + \tfrac{1}{2}Fs^2$$

(*energy* = *mass* × *acceleration* × *height* + $\tfrac{1}{2}$*force* × *displacement*$^2$)

The unit of energy is J which is $kg\,m^2s^{-2}$. The unit of *mgh* is $(kg)\,(m\,s^{-2})(m) = kg\,m^2s^{-2}$. So this part of the equation is fine.

The unit of $\tfrac{1}{2}Fs^2$ is $(kg\,m\,s^{-2})(m^2) = kg\,m^3s^{-2}$.

This is not the same as the unit for energy so the equation cannot be correct. **Note:** Even when the terms have the same unit it does not mean that the equation is definitely correct as there may be constants with or without units that have not been accounted for.

## Table of prefixes

When using units ensure you take account of prefixes and learn how to convert them to standard form.

*Table I.4*

| Prefix | Meaning | Multiplying factor |
|---|---|---|
| G | giga | $10^9$ |
| M | mega | $10^6$ |
| k | kilo | $10^3$ |
| c | centi | $10^{-2}$ |
| m | milli | $10^{-3}$ |
| μ | micro | $10^{-6}$ |
| n | nano | $10^{-9}$ |
| p | pico | $10^{-12}$ |
| f | femto | $10^{-15}$ |

### Test your understanding

1  Write down the following in SI base units:
   (a)  V (unit of potential difference)
   (b)  Pa (unit of pressure)
   (c)  Ω (unit of resistance).
2  The equation for a force $F$ acting on a charge $Q$ that is moving at a velocity $v$ in a magnetic field of flux density $B$ is $F = BQv$. Determine the unit for $B$ in terms of base units.
3  Check if the following equations could be correct.
   (a)  *time constant* = *resistance* × *capacitance*
        (*resistance* = *voltage/current* & *capacitance* = *charge/voltage*)
   (b)  *Work done* = *pressure* × (*volume*)$^2$

Candidates may find the following formulae useful when answering questions in AS and A2 assessment units.

**11**

Formulae

## Foundation physics mechanics formulae

$$\text{moment of force} = Fd$$

$$v = u + at$$

$$s = ut + \tfrac{1}{2}at^2$$

$$v^2 = u^2 + 2as$$

$$s = \tfrac{1}{2}(u + v)t$$

$$\text{for a spring } F = k\Delta l$$

$$\text{energy stored in a spring} = \tfrac{1}{2}F\Delta l = \tfrac{1}{2}k(\Delta l)^2$$

$$T = \frac{1}{f}$$

## Foundation physics electricity formulae

$$I = nAvq$$

$$\text{terminal p.d.} = E - Ir$$

$$\text{series circuit } R = R_1 + R_2 + R_3 + \dots$$

$$\text{parallel circuit } \frac{1}{R} = \frac{1}{R_1} + \frac{1}{R_2} + \frac{1}{R_3} + \dots$$

output voltage a

## Waves and nuclear physics formulae

$$\text{fringe spacing} = \frac{\lambda D}{d}$$

$$\text{single slit diffraction minimum } \sin\theta = \frac{\lambda}{b}$$

$$\text{diffraction grating } n\lambda = d\sin\theta$$

$$\text{Doppler shift } \frac{\Delta f}{f} = \frac{v}{c}$$

$$\text{for } v \ll c$$

$$\text{Hubble law } v = Hd$$

$$\text{radioactive decay } A = \lambda N$$

## Properties of quarks

| Type of quark | Charge | Baryon number |
|---|---|---|
| up u | $+\tfrac{2}{3}e$ | $+\tfrac{1}{3}$ |
| down d | $-\tfrac{1}{3}e$ | $+\tfrac{1}{3}$ |
| $\bar{\text{u}}$ | $-\tfrac{2}{3}e$ | $-\tfrac{1}{3}$ |
| $\bar{\text{d}}$ | $+\tfrac{1}{3}e$ | $-\tfrac{1}{3}$ |

## Lepton numbers

| | Lepton number L | | |
|---|---|---|---|
| e | $L_e$ | $L_\mu$ | $L_\tau$ |
| | 1 | | |
| | −1 | | |
| | 1 | | |
| | −1 | | |
| | | 1 | |
| | | −1 | |
| | | 1 | |
| | | −1 | |
| | | | 1 |
| | | | −1 |
| | | | 1 |
| $\bar{\nu}_\tau$ | | | −1 |

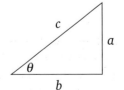

## Geometrical and trigonometrical relationships

$$\text{circumference of circle} = 2\pi r$$

$$\text{area of a circle} = \pi r^2$$

$$\text{surface area of sphere} = 4\pi r^2$$

$$\text{volume of sphere} = \tfrac{4}{3}\pi r^3$$

$$\sin\theta = \frac{a}{c}$$

$$\cos\theta = \frac{b}{c}$$

$$\tan\theta = \frac{a}{b}$$

$$c^2 = a^2 + b^2$$

## Circular motion and oscillations

$$v = r\omega$$

$$a = -\left(2\pi f\right)^2 x$$

$$x = A\cos 2\pi ft$$

$$\text{maximum } a = \left(2\pi f\right)^2 A$$

$$\text{maximum } v = 2\pi fA$$

$$\text{for a mass–spring system } T = 2\pi\sqrt{\frac{m}{k}}$$

$$\text{for a simple pendulum } T = 2\pi\sqrt{\frac{l}{g}}$$

## Fields and their applications

$$\text{For a uniform electric field, } E = \frac{F}{Q} = \frac{V}{d}$$

$$\text{For a radial field, } E = \frac{kQ}{r^2}$$

$$k = \frac{1}{4\pi\varepsilon_0}$$

$$g = \frac{F}{m}$$

$$g = \frac{GM}{r^2}$$

$$\text{for point masses, } \Delta E_p = GM_1M_2\left(\frac{1}{r_1} - \frac{1}{r_2}\right)$$

$$\text{for point charges, } \Delta E_p = kQ_1Q_2\left(\frac{1}{r_1} - \frac{1}{r_2}\right)$$

$$\text{for a straight wire } F = BIl$$

$$\text{for a moving charge } F = BQv$$

$$\phi = BA$$

$$\text{induced emf} = \frac{\Delta\left(N\phi\right)}{t}$$

$$E = mc^2$$

## Temperature and molecular kinetic theory

$$T/K = \frac{\left(pV\right)_T}{\left(pV\right)_{tr}} \times 273.16$$

$$pV = \frac{1}{3} Nm \left\langle c^2 \right\rangle$$

$$\text{energy of a molecule} = \frac{3}{2} kT$$

## Heating and working

$$\Delta U = Q + W$$

$$Q = mc\Delta\theta$$

$$Q = ml$$

$$P = Fv$$

$$\text{efficiency} = \frac{\text{useful power output}}{\text{power input}}$$

$$\text{work done on gas} = p\Delta V$$

$$\text{work done on a solid} = \frac{1}{2}F\Delta l$$

$$\text{stress} = \frac{F}{A}$$

$$\text{strain} = \frac{\Delta l}{l}$$

$$\text{Young's modulus} = \frac{\text{stress}}{\text{strain}}$$

## Capacitance and exponential change

$$\text{in series } \frac{1}{C} = \frac{1}{C_1} + \frac{1}{C_2}$$

$$\text{in parallel } C = C_1 + C_2$$

$$\text{energy stored by capacitor} = \frac{1}{2}QV$$

$$\text{parallel plate capacitance } C = \frac{\varepsilon_\circ\varepsilon_r A}{d}$$

$$Q = Q_\circ e^{-t/RC}$$

$$\text{time constant} = RC$$

$$\text{time to halve} = 0.69RC$$

$$N = N_\circ e^{-\lambda t}$$

$$A = A_\circ e^{-\lambda t}$$

$$\text{half-life } t_{\frac{1}{2}} = \frac{0.69}{\lambda}$$

## Momentum and quantum phenomena

$$\text{momentum} = mv$$

$$Ft = \Delta(mv)$$

$$E = hf$$

$$hf = \Phi + E_{k(\text{max})}$$

$$hf = E_2 - E_1$$

$$\lambda = \frac{h}{mv}$$

## Synoptic Assessment

In Units 5 and 6 you will be assessed on your ability to:
- bring together principles and concepts from different areas of physics and apply them in a particular context
- use the skills of physics in contexts that bring together the different areas of the subject.

Questions designed to test this ability are **synoptic questions**. There are three general ways in which you will be tested.

You will need to be able to apply physics in:
- practical work when creating hypotheses, justifying design and evaluating
- questions that are based on an application of physics
- questions based on a passage about some aspect of physics (a comprehension question).

The first two are basically the same. Questions will direct you to focus on particular aspects of an experiment or application.

EXAMPLE: To simulate the high accelerations experienced during take off, astronauts are trained using a system similar to the one shown in Figure I.1.

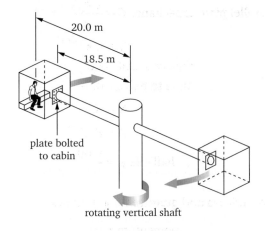

**Figure I.1**

The astronaut sits, as shown, and the system is accelerated to a high speed.

The centre of mass of the astronaut is 20.0 m from the axis of rotation.
(a) Explain why there is a horizontal force on the astronaut, even when the speed is constant.
(b) Write down the formula for the horizontal force on the astronaut, defining all the terms that are used.

(c) In one test, the astronaut experiences a horizontal force of 4.0 times that due to normal gravity. Calculate:
  (i) the speed of the astronaut during this test.
  (ii) the frequency of rotation in revolutions per second during the test.
  (iii) the magnitude and direction of the resultant force on the astronaut.
(d) The cabin is held on to the rotating arm by bolts. Each bolt has a cross-sectional area of $1.5 \times 10^{-4}$ m$^2$. The tensile stress in each bolt must be no greater than 80 MPa when in use. The total mass of the cabin, including the astronaut and equipment is $1.5 \times 10^3$ kg. The centre of mass is 18.5 m from the axis of rotation.

Assuming that the speed calculated in (ii) is the maximum required, calculate the minimum number of bolts needed.
See page 140 for answers.

## Comprehension

The best approach to a comprehension question depends on the individual. Different teachers may recommend different methods and you should find out which approach suits you.

The following is one suggested approach:
- skim-read the passage to get the gist of what it is about
- read through the questions to familiarise yourself with what aspects of the passage you need to concentrate on
- read the passage carefully, noting information that you identify as likely to be relevant when answering the questions (in the examination you may wish to underline or highlight the relevant information)
- answer the questions referring to those parts of the passage that you have identified, or to those parts that you are directed to by line references in the question
- answer the questions sequentially as some later questions may need answers from earlier parts.

When answering a question:
- do not copy chunks of the passage
- where you are asked to expand on an idea you should do this in your own words
- ensure that you always support any points you make in a way that demonstrates your understanding of the physics involved
- make clear which part of a question you are answering by using question and part numbers.

### A comprehension for you to try

Read this article and answer the questions that follow.

Modern microelectronic circuits are based on the silicon chip. Silicon exists naturally in sand, 92% being the stable isotope $^{28}_{14}$Si. For use in microchips, the silicon crystal used must contain no more than 1 in $10^9$ foreign atoms.

'Pure' silicon has a resistivity of $2.3 \times 10^3\ \Omega$ m. This is about $1 \times 10^{11}$ times greater than that of copper, and polythene has a resistivity $1 \times 10^{12}$ greater than that of silicon. The resistivity of the silicon is changed when other elements are introduced into the lattice structure in the process known as doping. Producing a pure silicon crystal is the first essential step in the production of a microchip. The method used is called zone refining. A diagram of the apparatus used in zone refining is shown in Figure I.2.

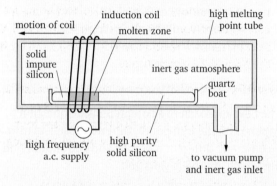

**Figure I.2**

The bar is placed in a quartz boat. The boat is placed in a chamber that is evacuated or filled with an inert gas such as argon. A high frequency current is passed through the coil and the boat is moved slowly through the coil.

The induced currents in the bar cause heating due to the resistance of the silicon, and the silicon (melting point 1420°C) melts.

As the silicon re-solidifies on leaving the coils, the impurities in the silicon remain in the molten part of the bar. When all the impurities have been swept to one end, this can be removed, and the pure silicon sliced into thin circular wafers that have an area of about $1.5 \times 10^{-3}$ m$^2$ and a thickness of about 1 mm.

In a process called 'doping', impurity atoms are introduced into the surface of the silicon crystal to form p and n type semiconductors. When these are manufactured next to each other they form a p–n junction.

One use of such a junction is as a solid-state detector for radiation from radioactive materials. One such detector is shown diagrammatically in Figure I.3.

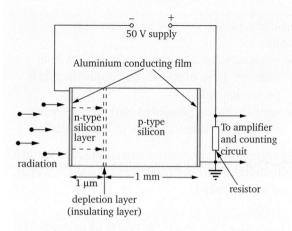

**Figure I.3**

The region between the p and n type materials contains no charge carriers. This is an insulating layer called a 'depletion layer'. A voltage, connected as shown, produces an electric field in the depletion layer. When ionising radiation passes through the depletion layer, the depletion layer becomes conducting. This results in a short voltage pulse across the resistor **R**.

### Test your understanding

1 Assuming that all the silicon in a crystal is silicon-28, calculate the number of impurity atoms that there can be in a 5.0g crystal of silicon that is pure enough to use in making microcircuits.
   (Avogadro constant $N_A = 6.0 \times 10^{23}$)

2 Use data from paragraph 2 on this page:
   (a) to determine the ratio of the resistivity of polythene to that of copper
   (b) to determine the resistance between the opposite faces of the wafer described in paragraph 5.

3 (a) Explain how the heating occurs in the bar when a high frequency current is generated in the coil.
   (b) Suggest why the method uses a high frequency current rather than one at a much lower frequency.
   (c) Explain one property that makes the quartz suitable for use as the container of the silicon.

4 Sketch a graph showing how the current through a diode varies when a potential difference is applied across it in each direction.

5 State the factors that affect the magnitude of the electric field in the depletion layer.

6 Explain how the voltage pulse is generated when an ionising particle enters the depletion layer.

See page 140 for answers.

## Unit 6 Experimental Skills

### The Practical Examination

The Unit 6 practical examination is in two parts.

#### Exercise 1 (1½ hours allowed)

This consists of one long question that tests your ability to **carry out**, and **analyse** data from, an experiment. The question may be set in any topic area from either of the AS or A2 courses.

You will be provided with some instructions, but these will be more limited than in the AS Unit 3 examination and you may need to select appropriate apparatus from a selection provided.

This exercise will be completed before the main examination period. Your school or college will advise you when this exercise is to be undertaken.

You will be provided with a **briefing sheet** some time before this exercise. You will therefore know the general area in which the activity is set.

#### Exercise 2 (1½ hours allowed)

This will be time-tabled to take place during the normal examination period. It consists of two tasks that test your ability to **plan**, **analyse** and **evaluate** experimental arrangements or procedures.

Each part of the exercise may require you to make some measurements.

Either of the two exercises may contain **synoptic questions**. In these questions you will be required to answer questions relating to physics ideas that are relevant. These may be from any part of the AS or A2 courses. There are likely to be more questions of this type in the second exercise than in the first.

You could, for example, be asked to explain how physics principles have been applied in the design of the apparatus, or to use physics from the AS and A2 courses to explain qualitatively why a particular term in a formula affects the outcome as it does.

### Experimental Skills

The four experimental skills areas are the same as those in the AS course:

- planning
- implementing
- analysing
- evaluating.

In A2 you will be expected to:
- demonstrate these skills in more difficult contexts and with less guidance
- demonstrate a greater maturity in experimental work than at AS level
- provide more detailed responses to questions that need explanations or descriptions than in the AS course, using your knowledge from both AS and A2
- be familiar with specific practical skills from both the AS and A2 course topics.

To remind yourself of the more basic practical skills, you should refer to pages 11 to 14 of the AS text book. These contain a detailed review of the four skill areas and include the basic work on analysing data with and without graphs.

In A2 you must be able to analyse data using logarithmic graphs. Only this additional skill is covered in this book.

### Logarithmic Graphs

Practical work usually has the objective of investigating whether an experimental relationship between two quantities agrees with a given hypothesis (or theory) or finding an equation that relates two quantities. The use of logarithms is a powerful tool in this task.

When the variables in an experiment produce a curve such as that in Figure I.4 when plotted directly on a graph, the variables may be related by a power law or by an exponential relationship. A logarithmic graph enables you to decide whether or not this is the case.

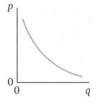

*Figure I.4*

## What is a Logarithm?

Logarithms are linked to indices.

Using index notation, $100 = 10^2$.

We say that 2 is the logarithm of 100 to the base 10.

The logarithm of a number tells us to what power the base has to be raised to give the number itself.

So: 10 has to be raised to the power 2 to get 100
10 has to be raised to the power 3 to get 1000, etc.

### Base 10 and base e

These are the two important bases for scientific work. Logarithms to the base e are called **natural logarithms**.

The value of e is 2.718 to 4 significant figures. (e is a 'never-ending' number like $\pi$)

### Calculating log and ln of a number

When you see **log** it means the base is 10. When you see **ln** it means the base is e.

These are the labels on the keys on your calculator. To determine the log or ln simply insert the number and press log or ln.

Check that you can use your calculator by confirming the log and ln of the following numbers.

**Table I.5**

|  | log | ln |
|---|---|---|
| 3.54 | 0.549 | 1.264 |
| 0.000267 | $-3.573$ | $-8.228$ |
| $2.45 \times 10^{14}$ | 14.389 | 33.132 |
| $1.73 \times 10^{-8}$ | $-7.762$ | $-17.873$ |

When you use logarithms, do not include the number before the decimal point in your significant figure assessment. This will cause you to lose accuracy in your calculations and graphs.

### Antilog (or inverse log)

Finding the antilog is the process of calculating the actual number that corresponds to a given log or ln.

### For base 10

This is called the antilog (INV Log on the calculator).

If $y$ is the log, the antilog is simply the value of $10^y$. From the table, the antilog of 0.549 is 3.54.

### For base e

If $y$ is the ln, the number is $e^y$. From the table the INV ln of 33.132 is $2.45 \times 10^{14}$.

### Some useful relationships when using logarithms

$\ln\ e = 1$ $\qquad\qquad$ $\log\ 10 = 1$

**For ln**

$$\ln\ xy = \ln\ x + \ln\ y$$

$$\ln\ (x/y) = \ln\ x - \ln\ y$$

$$\ln\ x^n = n \ln\ x$$

Similar relationships hold for logs.

### Logarithms and units

You can only find the logarithm of a pure number so **a logarithm has no unit**.

To determine the log or ln of a quantity, the quantity first has to be divided by its unit to convert it to a pure number.

For example, if you wish to determine the log of a length $l$ measured in m you determine the log of $(l/m)$. You should therefore write **log** $(l/m)$ at the head of a column in a table of data or on the axis of a graph.

## Graphical Analysis using Logarithms

Note:
- when plotting logarithmic graphs you will usually need to include a false origin. If you do not use a false origin the points will usually be crammed into only a small part of the $x$ or $y$ scale.
- the log of a number that is less than 1 is negative, so take care when you are plotting the graph.

### To test whether a quantity $q$ varies exponentially with $p$

1 Determine the log or ln of the quantities for log $q$.
2 Do nothing to the values for $p$.
3 Plot a graph of log $q$ or ln $q$ against $p$. That is plot log $q$ or ln $q$ on the y axis and $p$ on the x axis.

If the best line through the points is a straight line then $q$ varies exponentially with $p$.

A negative gradient shows that the change is an exponential decay.

A positive gradient shows that the change is an exponential growth.

## Mathematical reasoning

The equation for **an exponential decay** of a quantity $n$ with time $t$ is

$$n = n_o e^{-kt}$$

Where $n$ is the quantity that is changing with time $t$
$n_o$ is the value of $n$ at time $t = 0$ (since $e^0 = 1$)
$k$ is a constant for the decay.

Taking ln (logarithms to the base $e$) on both sides gives:

$$\ln n = \ln n_o + \ln (e^{-kt})$$

$$\ln n = \ln n_o - kt$$

$$\text{so, } \ln n = -kt + \ln n_o$$

Now compare the ln equation with the general equation for a straight line.

$$\begin{array}{ccccc} \ln n & = & -kt & + & \ln n_o \\ \Downarrow & & \Downarrow \Downarrow & & \Downarrow \\ y & = & mx & + & c \end{array}$$

A plot of $\ln n$ on the $y$ axis and $t$ on the $x$ axis gives a graph like that in Figure I.5.

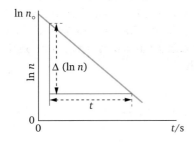

**Figure I.5**

- The gradient $\left(\dfrac{\Delta(\ln n)}{\Delta t}\right)$ of the graph is $-k$.

- Since the gradient is negative, $k$ is a positive number.
- The unit for $k$, the gradient, is $s^{-1}$.
- The intercept on the ln $n$ axis is $\ln n_o$.
- Calculating INV ln of the intercept gives $n_o$.

$n$ could represent the number of radioactive atoms present when a radioactive substance decays.

During the work on capacitors, you will investigate how the pd across a capacitor, the charge on a capacitor and the current in the circuit vary with time during discharge. In this case the quantity $n$ in the above equations is replaced by either $V$, $Q$ or $I$.

The exponential change is often a variation with time but this is not essential. It could, for example, be a distance as in the following question.

## Test your understanding

The table shows the intensity $I$ of gamma radiation that passes through absorbers of different thickness $t$ as a percentage of the original intensity.

**Table I.6**

| $t$/m | $I$/% |
|---|---|
| 0 | 100 |
| 0.020 | 73 |
| 0.040 | 54 |
| 0.060 | 40 |
| 0.080 | 29 |
| 0.100 | 21 |
| 0.120 | 16 |

Theory suggests that the equation for the intensity is $I = I_o e^{-\mu t}$ where $t$ is the thickness of the absorber.
(a) Show that the data agree with this theory.
(b) Determine the value for $\mu$ for the absorber used.
(c) Determine the thickness that reduces the intensity to half the original value.

## Can log be used instead of ln?

Yes it can. The graph will still be linear for an exponential change, but the gradient and intercept have different meanings.

For a graph of log $q$ against $p$, the gradient will be:

$$\frac{\text{the decay constant}}{2.30} = \frac{-k}{2.30} \text{ and}$$

$$\text{the intercept} = \log n_o$$

## Power Laws

The power laws that are common in the A2 course are the inverse and inverse square laws such as those involved in work on electric fields.

$$V \propto \frac{1}{r} \text{ and } E \propto \frac{1}{r^2}$$

These may be written as:

$$V \propto r^{-1} \text{ and } E \propto r^{-2}$$

A power relationship between two quantities $p$ and $q$ always has the form:

$$p = A q^n$$

where $A$ and $n$ (the power) are constants. The constant $A$ may be a positive or a negative number but in practical situations is usually positive.

The constant $n$ may also be positive or negative and may be a fraction or a decimal. In many laws of physics the value of $n$ is a number such as $\pm\frac{1}{2}, \pm 1, \pm 2$ but the number may be any value in practice.

### To test whether *q* is related to *p* by a power law

**1** Determine the log or ln of **both** *p* and *q*.
Do not take the ln of one and the log of the other. You must do the same for both.

**2** Plot a graph of log *q* against log *p* or ln *q* against ln *p* (i.e. plot log *q* or ln *q* on the *y* axis and log *p* or ln *p* on the *x* axis).

If the best line through the points is a straight line then the relationship between the two quantities is a power law.

A negative gradient shows an inverse proportionality.

$$q \propto \frac{1}{p^n}$$

A positive gradient shows direct proportionality.

$$q \propto p^n.$$

### Mathematical reasoning

Using some of the useful relationships on page 16 and taking logarithms of both sides of the equation $p = A q^n$ gives:

$$\log p = \log (A q^n)$$
$$\log p = \log A + \log (q^n)$$
$$\log p = \log A + n \log q$$

Rearranging this gives:

$$\log p = n \log q + \log A$$

Note that since *A* is constant its logarithm is also constant.

## Determining the Constants in a Power Law

Plot the graph using either ln or log values. It is assumed here that the graph is the log graph.

Compare the log form of the equation with the general equation for a straight line.

$$\log p = n \quad \log q + \log A$$
$$\Downarrow \quad \Downarrow \quad \Downarrow \quad \Downarrow$$
$$y = m \quad x \ + \ c$$

log *p* is plotted on the y axis.
log *q* is plotted on the x axis.

The graph in Figure I.6 is for a situation in which *n* is positive.

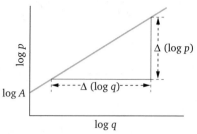

**Figure I.6**

- The **gradient** of the graph gives the value of **n** and is given by:

$$n = \frac{\text{change in log } p}{\text{change in log } q} = \frac{\Delta(\log p)}{\Delta(\log q)}$$

- The logs do not have units so the gradient has no unit.
- The **intercept on the log *p* axis is log *A*.**
- Calculating the INV log of the intercept gives *A*.

**Take care to notice whether or not there is a false origin when determining the intercept. A false origin is common in this type of graph.**

EXAMPLE: The following data was obtained in an experiment to investigate how the count rate (corrected for background radiation) *C* varied with distance *d* from a radioactive source.

**Table I.7**

| *d*/cm | 3.0 | 4.0 | 5.0 | 6.0 | 7.0 |
|---|---|---|---|---|---|
| *C*/s$^{-1}$ | 56 | 32 | 20 | 14 | 10 |

The count rate is expected to obey a power law.

**(a)** Show that the data is consistent with a power law.

**(b)** Determine the equation that relates count rate *C* to distance *d* from the source.

**1 First find the logs of *C* and *d*.**

**Table I.8**

| log (*d*/cm) | 0.477 | 0.602 | 0.700 | 0.778 | 0.845 |
|---|---|---|---|---|---|
| log (*C*/s$^{-1}$) | 1.78 | 1.51 | 1.30 | 1.15 | 1.00 |

## 2 Next plot the log graph.

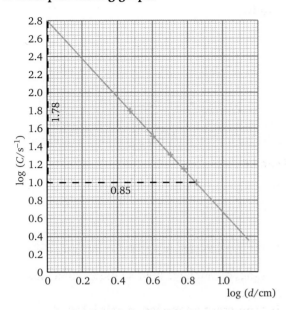

## 3 Finally, interpret the graph.

The best line through the points is a straight line with a negative gradient. The data are related by an inverse power law. The graph should have the form $C = A\, d^n$ where $n$ will be negative. The equivalent log equation is:

$$\log\ C = \log\ A + n \log\ d$$

The gradient gives $n$. This is $-1.78/0.85 = -2.1$ so the law is $C \propto 1/d^{2.1}$.

The intercept on the log $(C/s^{-1})$ axis is 2.78.

This is log $A$.

The value for the $A$ is INV log 2.78. $A$ is therefore 603 $s^{-1}$.

The equation relating $C$ and $d$ from this data is:

$$C = 603d^{-2.1} = \frac{603}{d^{2.1}} \text{ or rounding off } C = \frac{600}{d^2}$$

(The constant 2.1 is close to 2 so an inverse square law is a distinct possibility.)

## Test your understanding

1 The following data were obtained for the variation of the frequency $f$ of oscillation of the oscillating mass $m$.

**Table I.9**

| $f$/Hz | 32 | 23 | 16 | 13 | 11 |
|---|---|---|---|---|---|
| $m$/kg | 0.050 | 0.100 | 0.200 | 0.300 | 0.400 |

The equation for the oscillator is $f = km^n$.
   (a) Showing any mathematical reasoning, determine the values of $k$ and $n$.
   (b) Determine the frequency when $m = 0.150$ kg.
   (c) Determine the mass that gives a frequency of 25 Hz.

2 The following data relate the radius $R$ of a nucleus to the nucleon number $A$ of the nucleus. The relation between them is thought to obey the law:

$$R = R_o A^n$$

Analyse the data to check that this is the case and determine the values of $R_o$ and $n$.

**Table I.10**

| $R/10^{-15}$m | 3.2 | 5.3 | 6.0 | 7.2 | 8.0 |
|---|---|---|---|---|---|
| $A$ | | 12 | 56 | 81 | 138 | 184 |
| Element | | C | Fe | Br | Ba | W |

# Circular Motion

## Measuring Angles

To study circular motion you need to be familiar with the units that are in regular use when measuring angles and angular speeds, and the relationships between them.

Angles can be measured in degrees or radian (rad).

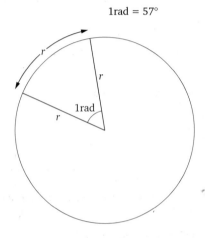

1rad = 57°

**Figure 4.1**

**One radian** is the angle subtended at the centre of a circle by an arc of length equal to the radius (see Figure 4.1).

$$\text{angle in radian} = \frac{\text{arc length}}{\text{radius}}$$

Since the total arc length in a complete circle is $2\pi r$ the angle that corresponds to a complete circle is $2\pi r/r$, therefore:

$$2\pi \text{ rad} = 360°$$

## Angular Speed (Angular Velocity)

Rotational speeds (such as the speed of a motor) are often given in **revolutions per second** or revolutions per minute. This is the frequency $f$ of the revolutions.

In scientific work it is more useful to give the angular velocity in **radian per second** (rad s$^{-1}$).

This is essential when using the equations relating to circular motion.

The angular velocity is the angle that a radius arm moves through in 1 second when an object completes $f$ revolutions per second, therefore:

$$\text{angular velocity } \omega = 2\pi f \text{ or } \frac{2\pi}{T}$$

where $T$ is the period, the time for one revolution.

### Relationship between linear and angular velocities

When an object is moving in a circular path at constant speed it completes one revolution in a given time $T$.

The velocity $v$ at any instant is given by $v = \dfrac{2\pi r}{T}$

where $r$ is the radius of the circular path. The direction of the velocity is at a tangent to the circular path.

Since $\omega = \dfrac{2\pi}{T}$ it follows that $v = r\omega$.

## Central (Centripetal) Force and Acceleration

**Acceleration** is rate of change of **velocity** and velocity has magnitude (speed) and direction. Acceleration can therefore be produced either by a change in speed or a change in direction (or both).

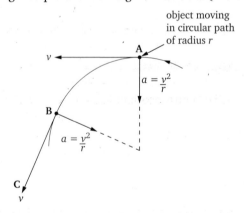

**Figure 4.2**

During circular motion, although the linear speed is constant:
- the **direction** is continually changing
- so the **velocity** is continually changing
- so there is an **acceleration**

- the **direction of the acceleration** is always toward the centre of the circular path
- a **force** is needed toward the centre of the circular path to produce the acceleration.

This is illustrated in Figure 4.2. The velocity of the object is always at a tangent to the circular path so when the object moves from **A** to **B** the velocity has changed.

The acceleration and force toward the centre of a circular path are referred to as central (or centripetal) acceleration and force.

The **magnitude of the acceleration** $a$ of the object is given by:

$$a = \frac{v^2}{r} \text{ or } r\omega^2$$

Using Newton's law, $F = ma$, it follows that the magnitude of the central force is given by:

$$F = \frac{mv^2}{r} \text{ or } mr\omega^2$$

You do not need to recall the mathematical derivation of these formulae for examination purposes so it is not included here. To improve your understanding, you may wish to look up the derivation, which will be found in many physics text books.

EXAMPLE: A disc of radius 0.25 m is spinning at a speed of 3000 revolutions per minute.
  (a) Calculate:
    (i) the angular speed of the disc in rad s$^{-1}$
    (ii) the linear speed of a particle at the rim of the disc
    (iii) the central acceleration of the particle.
  (b) The particle has a mass of $1.5 \times 10^{-25}$ kg. Calculate the central force acting on the particle.
  (a) (i) Angular speed $\omega = 2\pi f = 2\pi(3000/60)$
           $= 314$ rad s$^{-1}$
      (ii) Linear speed $= r\omega = 0.25 \times 314$
           $= 78.5$ m s$^{-1}$
      (iii) Central acceleration $= v^2/r = 78.5^2/0.25$
             $= 2.46 \times 10^4$ m s$^{-2}$
  (b) Central force $= m \times$ central acceleration
         $= 3.7 \times 10^{-21}$ N

## Central forces in practice

Whenever an object moves in a circular path you should be able to identify how the central force is produced.

The force may be produced by friction, reaction forces or tension in rods or wires. You will study circular motion involving gravitational, electrical and magnetic forces in Module 5.

A few practical situations are shown in Figure 4.3. The origin of the central force $F$ and the centre **C** of the circular path are shown in each diagram.

(a)

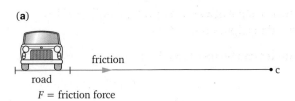

$F = $ friction force

**Car going around a bend on a level road**

(b)

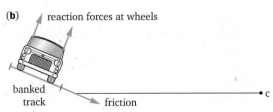

$F = $ Sum of horizontal components of friction and reaction forces

**Car going round a banked track**

(c)

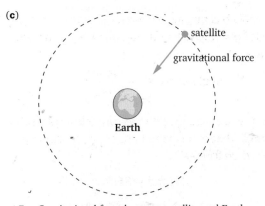

$F = $ Gravitational force between satellite and Earth

**Satellite in an orbit around Earth**

**Figure 4.3**

## What happens if the force disappears?

At the instant the force ceases to exist, the object will move in a straight line at constant speed. The path will be the tangent to the circular path at the instant the force disappears. For example, in **Figure 4.2,** if the force suddenly disappeared when the object was at position **B** the object would move in the direction **BC**.

**Test your understanding**

**1** **(a)** A car of mass 800 kg is travelling at a constant speed of 10 m s$^{-1}$ around a bend on a flat road. The radius of the bend is 50 m. Calculate the central force acting on the car.

**(b)** The maximum frictional force between the tyres and the road is 0.4 times the weight of the car. Calculate the maximum speed that the car can travel round the bend without skidding. ($g = 9.8$ m s$^{-2}$)

**2** A mass of 0.20 kg is attached to one end of a string that breaks when the force applied is 40 N. The mass is whirled round at a constant angular speed of 10 rad s$^{-1}$. The length of string is gradually increased.

**(a)** Calculate:

**(i)** the frequency of rotation in revolutions per second

**(ii)** the period of rotation

**(iii)** the radius of the path when the string breaks.

**(b)** Sketch a diagram to show the circular path and the direction of motion when the string breaks.

**3** **(a)** Explain why a particle is accelerating even when it is moving at a uniform speed in a circular path.

**(b)** In a synchrotron, protons of mass $1.7 \times 10^{-27}$ kg are accelerated to a high speed. They move in a circular path that has a constant radius of 400 m. They initially have a speed of $8.0 \times 10^6$ m s$^{-1}$.

**(i)** Calculate the magnitude of the central acceleration of the proton.

**(ii)** Calculate the magnitude of the central force needed to produce the circular motion.

**(iii)** Draw a graph showing how the magnitude of the central force would vary for proton speeds in the range 0 to $20 \times 10^6$ m s$^{-1}$.

**4** **(a)** Figure 4.4 shows a body moving with uniform speed in a horizontal circle.

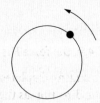

**Figure 4.4**

**(i)** Copy the diagram and show on it the direction of the resultant force $P$ acting on the body.

**(ii)** Show in your diagram the path the body would follow if the force ceases to exist when it is at the point shown. Label this path **D**.

**(b)** In a fairground ride called a 'rotor', a person of mass 60 kg stands against a wall, as shown in Figure 4.5, and the wall is rotated. When it is spinning at a suitable speed the floor is dropped so that the person is left 'stuck to the wall'.

Figure 4.6 shows the variation of frictional force $F$ with normal reaction between the person and the wall.

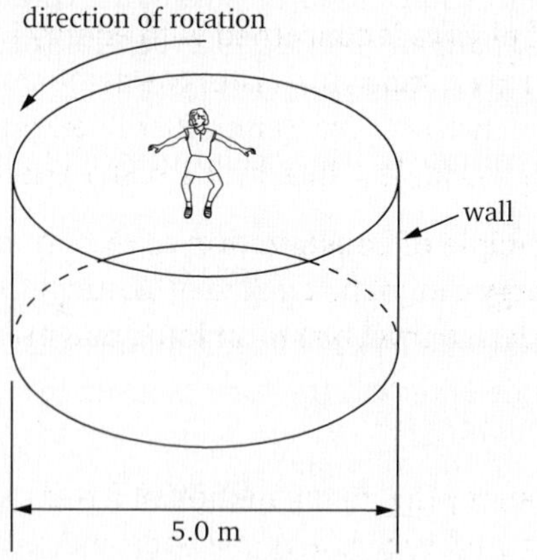

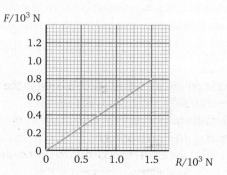

**Figure 4.5**

**Figure 4.6**

Determine:

**(i)** the normal reaction when the frictional force is equal to the weight of a person of mass 60 kg.

**(ii)** the minimum angular speed, in rad s$^{-1}$, at which such a person must be rotated to remain in position when the floor is dropped.

(Gravitational field strength = 9.8 N kg$^{-1}$.)

# Work, Energy and Power

## Energy

Much of physics is concerned with energy in its various forms and the transfer of energy from one form into another. The SI unit for all forms of energy is the joule J. Energy is a scalar quantity.

The **principle of conservation of energy** states that energy cannot be created or destroyed but can be transferred from one form into another.

### Kinetic energy

This is energy due to the motion of a body and depends on the mass $m$ of the body and its speed $v$.

$$E_K = \tfrac{1}{2}mv^2$$

### Potential energy

This is energy due to the position of the body. It may be:

- elastic potential energy (such as due to the stretching of a spring or wire)
- gravitational potential energy due to the attraction of two masses
- electrical potential energy due to the attraction (or repulsion) of two charged objects.

Gravitational and electrical potential energy are studied in detail in Unit 5. Only knowledge of the simplified formula for change in gravitational PE, $\Delta E_p$, is needed in this unit.

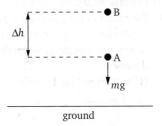

**Figure 4.7**

For the situation where the object moves from **A** to **B** in Figure 4.7, the change in gravitational potential energy is given by:

$$\Delta E_p = mg\Delta h$$

where   $m$ is the mass of the body
$g$ is the gravitational field strength
$\Delta h$ is the change in height (i.e. the change in separation of the two masses).

Other energy forms and transformations will be discussed later in this unit.

## Work

Mechanical work is done when a force $F$ is applied to an object and there is a component of the movement of the object **in the direction of the force**.

For example, when an object falls there is a force due to gravity. Work is done as the gravitational force on the object moves downwards. The potential energy decreases and the kinetic energy increases.

The work done must be equal to the energy transferred. The unit of work is the joule $J$, the same as that for energy.

The work done may:

- result in a change in the kinetic energy of the object as a whole (as when an object falls)
- result in a change in the internal energy of the object and the surroundings, producing a change in temperature.

The second of these situations occurs:

- when an object is pushed against a frictional force
- when a gas is compressed in a piston
- when a wire is stretched or compressed.

For work done when compressing gases and stretching wires see pages 53 and 58.

### Calculating work done

In Figures 4.8 and 4.9, the force $F$ is constant.

In Figure 4.8, the resulting displacement $s$ of the object is in the same direction as the force. The work done is given by:

$$W = Fs$$

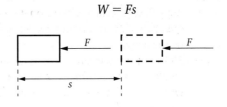

**Figure 4.8**

In Figure 4.9, the direction of the force $F$ and the displacement of the object $s$ are not in the same direction.

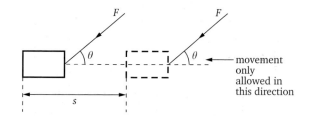

**Figure 4.9**

To calculate the work done you need to either:
- find the component of the force in the direction of movement ($= F \cos \theta$), or
- find the component of the displacement in the direction of the force ($= s \cos \theta$).

The result for work done is the same in each case and is given by:

$$W = Fs \cos \theta$$

In the situations in Figures 4.8 and 4.9:
- the system (a person or machine) that is applying the force and doing the work loses energy
- the object on which the force acts has work done on it and gains an equal amount of energy.

The energy that is transferred may be in the form of kinetic energy of the object and, if there are frictional forces involved, internal energy of the object and the surroundings.

EXAMPLE: In Figure 4.10, the tension in a rope that is pulling a truck along a straight rail-track is 2000 N.

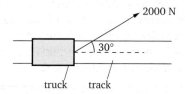

**Figure 4.10**

(a) Calculate the change in kinetic energy when the truck moves 3.0 m along the track assuming that there is no friction.

(b) The truck moves at a constant speed of 1.5 m s$^{-1}$ along the track.

    (i) Calculate the work done per second.

    (ii) What happens to the energy transferred when it is moving at a constant speed?

(a) Energy transferred   $= Fs \cos \theta$
                                 $= 2000 \times 3.0 \times \cos 30$
                                 $= 5200$ J

(b) (i)   Work done per second
                              $= 2000 \times 1.5 \times \cos 30$
                              $= 2600$ J

    (ii)   Since the truck is moving at constant speed there is no change in kinetic energy so all the energy is transferred into internal energy in the truck and the surrounding air.

## Work done by varying forces

In many practical situations the force is not constant. This is the case when stretching a spring. A different approach is needed to determine the work done.

### Spring stiffness

For a spring, the graph of applied force against the extension produced by the force is a straight line through the origin, as shown in Figure 4.11.

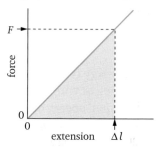

**Figure 4.11**

The force is proportional to the extension and the spring is said to obey **Hooke's law**. Some springs are harder to stretch (are stiffer) than others.

The **stiffness** (or **spring constant**) $k$ is a measure of how difficult it is to extend the spring and is defined by the equation:

$$k = \frac{F}{\Delta l} \text{ or } F = k \Delta l$$

where $F$ is the stretching force and $\Delta l$ is the extension it produces. The stiffness $k$ is the gradient of the graph in Figure 4.11.

### Energy stored in a spring

When a wire is stretched, a force moves its point of action. Therefore work is done on the wire and energy is transferred. The system that is doing the stretching loses energy and this becomes **elastic potential energy** in the spring. Provided that a spring has not been over-stretched, when the stretching force is removed the spring returns to its original size and all the stored energy is retrieved.

Initially only a small force is needed to produce an extension of 1 mm. As the spring is stretched, a greater force is required to produce further extension and more work has to be done to produce an extension of 1 mm.

When the force is varying with distance the work done is found from the area under the force – extension graph.

From Figure 4.11 the energy stored is the area of the shaded triangle and is given by:

$$\text{Elastic PE, } E = \tfrac{1}{2}F\Delta l$$

where $F$ is the stretching force and $\Delta l$ is the extension.

Since the spring stiffness $k$ is $\dfrac{F}{\Delta l}$ the stored elastic energy is also given by

$$E = \tfrac{1}{2}k(\Delta l)^2$$

In other situations, such as when a tennis ball is struck by a tennis racquet, the force on the ball increases from zero to a maximum and back to zero as shown in Figure 4.12.

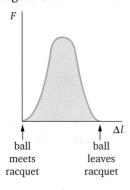

**Figure 4.12**

The work done is still the area under the force – distance graph between the graph line and the distance axis. This can be determined by:

- counting the number of squares under the curve
- calculating the work done that is represented by one square
- multiplying these together.

EXAMPLE ONE: A force of 5.0 N extends a spring by 0.015 m. Calculate the elastic energy stored in a spring:
- (a) when a 5.0 N force is applied
- (b) when a 7.5 N force is applied.

(a) Energy $= \tfrac{1}{2}F\Delta l = \tfrac{1}{2}5.0 \times 0.015 = 0.038$ J

(b) $k = \dfrac{5.0}{0.015} = 333 \text{ N m}^{-1}$

New extension $= \dfrac{7.5}{333} = 0.0225$ m

$E = \tfrac{1}{2} \times 333 \times 0.0225^2 = 0.084$ J

EXAMPLE TWO: The force acting on a ball when struck by a bat varies as shown in Figure 4.13.

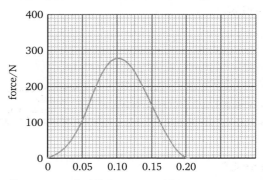

distance moved while in contact with racquet/m

**Figure 4.13**

Determine the kinetic energy of the ball as it leaves the bat.

Number of squares under the graph line ≈ 23.
Energy per square = 1.25 J.
Total energy given to the ball ≈ 29 J.

**Test your understanding**

1  Calculate the work done when a force of 150 N pushes an object 4.0 m:
   (a) in the direction of the force
   (b) in a direction at 40° to the direction of the force.
2  Calculate the work done when lifting a mass of 50 kg through a height of 1.3 m. (g = 9.8 N kg$^{-1}$)
3  The energy that was converted into potential energy when a climber of mass 75 kg climbed a rock face was 40 kJ. How high was the rock-face?
4  A spring that has a stiffness of 25 N m$^{-1}$ is compressed by 0.055 m.
   (a) Calculate the elastic energy stored.
   (b) Assuming that all this energy is changed to kinetic energy of an object of mass 0.20 kg, determine the speed of the object.
5  Figure 4.14 shows the force–extension graph when a rubber band was stretched.

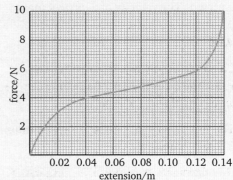

extension/m

**Figure 4.14**

   (a) Determine the work done when the band was stretched by 0.10 m.
   (b) The energy that is retrieved when the band contracts is less than that used when stretching the band. What becomes of the remaining energy?

## Power

Power $P$ is:
- the rate at which work is done
- the rate at which energy is transferred.

The unit of power is the watt W where $1\,\text{W} = 1\,\text{J}\,\text{s}^{-1}$: When work $Fs$ is done in time $t$:

$$P = \frac{Fs}{t} = Fv$$

where $v$ is the velocity $\left(\dfrac{s}{t}\right)$ in the direction of the force.

This formula gives the instantaneous power. Where the velocity is changed linearly (from $v_1$ to $v_2$) with a constant force acting, the average power can be found by using the average velocity, $\left(\dfrac{v_1 + v_2}{2}\right)$ in the above equation.

### Measuring your own power

1 Find the time $t$ that you take to run up a flight of stairs of known or measured vertical height $\Delta h$. The mean power developed is:
$$\frac{mg\Delta h}{t}$$

2 Measure the time $t$ that it takes you to lift a mass of a kg through a measured height $\Delta h$ (of about 1 m) a known number of times $n$. Your mean lifting power is:
$$\frac{nmg\Delta h}{t}$$

EXAMPLE ONE: The tractive force acting on a car travelling at $15\,\text{m}\,\text{s}^{-1}$ is 200 N. Calculate the useful power developed.

$$\text{Power} = Fv = 200 \times 15 = 3000\,\text{W}$$

EXAMPLE TWO: The velocity of a lorry of mass 7000 kg increases from $10\,\text{m}\,\text{s}^{-1}$ to $15\,\text{m}\,\text{s}^{-1}$ in 7.0 s.
Calculate:
(a) the change in kinetic energy of the lorry
(b) the average useful power developed by the lorry
(c) the force accelerating the lorry, assuming that it is constant.

(a) Change in KE $= \Delta(\tfrac{1}{2}mv^2)$
$= \tfrac{1}{2} \times 7000\,(15^2 - 10^2)$
$= 437\,\text{kJ}$

(b) Average power $= \dfrac{\Delta E_k}{time} = \dfrac{4.3 \times 10^5}{7.0} = 62.5\,\text{kW}$

(c) Average power = force × average velocity
Force $= \dfrac{62\,500}{12.5} = 5000\,\text{N}$
(Note that the force can be arrived at another way, by calculating the acceleration and using $F = ma$.)

## Efficiency

Practical devices such as motors, generators and engines are designed to convert energy from one form into another that is more useful.

Ideally all the energy input to the device would be converted into useful energy and the device would be 100% efficient. In practice, however, this is never the case. The efficiency is defined by

$$\text{Efficiency} = \frac{\text{useful power output}}{\text{power input}}$$

If the efficiency of an electric motor is 0.4 it means that 4/10 of the electrical power input becomes useful output of the motor. The energy transfers are shown diagrammatically in Figure 4.15.

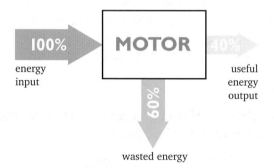

**Figure 4.15**

Efficiency is often quoted as a percentage so an efficiency of 0.4 is equivalent to an efficiency of 40%.

$$\text{Efficiency \%} = \frac{\text{useful power output}}{\text{power input}} \times 100\%$$

## Causes of Inefficiency

### Friction

In any machine with surfaces that rub against one another work has to be done against the frictional forces between the two surfaces. The energy transferred in this way is useless as far as the purpose of the machine is concerned. It is wasted energy and appears as internal energy in parts of the machine, producing a rise in temperature. (See page 51.)

To red  such losses, some practical devices use oil or othe  aterials to reduce friction between the two sur  es. Alternatively, ball bearings, which roll rather t  slide, may be used, and in very low friction  ces, an air interface is used between the two surf  s.

## Air (or f  ) resistance

A moving  rt of a machine may pass through air (or anoth  luid). The air has to be pushed out of the way and so work is done to transfer energy to the air as kinetic energy. The energy ultimately ends up producing a rise in internal energy of the air.

A force is needed to change the momentum of the air as it is pushed out of the way. This is exerted by the moving part of the machine, which itself experiences a retarding force. This force is referred to as air resistance.

## Resistive losses

In an electric motor there is current in the moving coil and in the coils producing the magnets. Although the resistance of the wire used is low it is not zero, so that electrical energy is transferred to internal energy in the wires. The energy loss is given by $I^2R$ and is wasted energy.

## Experiment to determine motor efficiency

The efficiency of an electric motor can be measured using the apparatus in Figure 4.16.

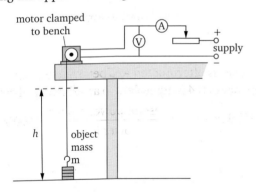

**Figure 4.16**

The motor is used to lift a suitable object of mass $m$. The mass is chosen so that the time $t$ taken to lift the object through a measured height $h$ (about 1 m) is easily measurable using a stop watch or timing gates (a few seconds). The object should move at a fairly constant speed.

The mean useful output power is given by $\dfrac{mgh}{t}$.

The input power to the motor is found by measuring the supply voltage and current while the object is being lifted. The input power is $VI$.

The efficiency as a percentage is then:

$$\frac{\text{useful output power}}{\text{input power}} = \frac{mgh/t}{VI} \times 100\%$$

### Test your understanding

1  Calculate the power generated by a weightlifter who lifts a 150 kg mass through a height of 1.2 m in 0.80 s. ($g = 9.8\ \text{N kg}^{-1}$)

2  A sprinter has an average output power of 700 W. A 100 m sprint takes 11.5 s.
   (a) Calculate the average force that the sprinter provides during the race.
   (b) Explain why this force does not continue to accelerate the sprinter.

3  The efficiency of a motor is 35%.
   (a) Calculate the useful output power when the input power is 500 W.
   (b) Calculate the input power when the useful output power is 300 W.
   (c) Calculate the energy wasted in 5.0 minutes when the input power is 500 W.
   (d) The output of this motor is fed to a gearing system that has an efficiency of 70%. Calculate:
      (i)  the efficiency of the complete system
      (ii) the power lost in the gears when the input to the motor is 500 W.

4  A person of mass 55 kg does 30 press-ups in one minute. The height the body rises during each press-up is 0.35 m. Calculate the power output of the person. ($g = 9.8\ \text{N kg}^{-1}$)

5  A car of mass 900 kg travels along a horizontal track at a constant speed of 25 m s$^{-1}$. The total resistive force acting is 1500 N. ($g = 9.8\ \text{N kg}^{-1}$)
   (a) Calculate the useful power output of the car on the horizontal track.
   (b) Calculate the useful power output needed to move at the same speed up a hill with a gradient of 20°.
   (c) The efficiency of the engine and transmission system of the car is 40%. Determine the power input to the car when moving on the horizontal track.

# Oscillations

## The Importance of Oscillators

The study of oscillations is important because of the wide range of applications in which they occur. The following is a list of some of the important applications.

- All waves can be traced back to something that is oscillating.
- Music is produced by vibrations in strings and pipes.
- Speech is produced by oscillations of vocal chords and cavities in the chest and throat.
- In the transmission of mechanical waves such as sound, the particles of the transmitting medium are oscillating.
- Oscillations of electrons in aerials produce electromagnetic waves.
- Parts of structures such as bridges, buildings, cars and washing machines are able to oscillate.

Oscillations are therefore essential in some applications and a disadvantage in others.

In the AS course you will have learned how to investigate oscillators. In this section you will study how those oscillators that move with **simple harmonic motion** can be modelled mathematically.

## Useful Definitions

*Frequency f* is the number of complete oscillations per second.
*Period T* is the time for one complete oscillation.

$$T = \frac{1}{f}$$

*Displacement x* is the distance between the position of the object and its equilibrium position. The displacement is a vector quantity as it can be in one of two directions.
*Amplitude A* is the maximum displacement of an oscillating object from its equilibrium position.
*The equilibrium position* is the point of zero displacement. It is where the oscillator will come to rest when it loses all its energy.

## Free oscillations

When a simple pendulum is displaced and released the bob oscillates with simple harmonic motion.

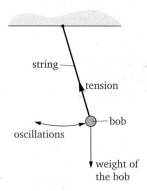

**Figure 4.17**

The only forces acting in this system are internal to the oscillator: the weight of the bob and the tension in the string. There are no external forces driving it.

The oscillation that occurs is called a **free oscillation**. The frequency of the oscillation that results in this case is called the **natural frequency of oscillation**.

The graph of the displacement against time is shown in Figure 4.18.

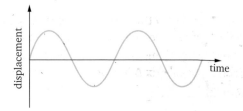

**Figure 4.18**

Similar displacement–time graphs are obtained when the masses in the mass–spring systems shown in Figure 4.19 are displaced and released.

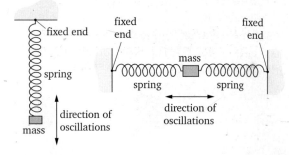

**Figure 4.19**

# Simple Harmonic Motion

When a body oscillates with **simple harmonic motion (shm)** the period of the oscillations is the same whatever the amplitude of the oscillation. **The period is independent of the amplitude.**

When an oscillating mass undergoes shm the displacement–time graph always has the same characteristic shape of a sine or cosine graph.

## Conditions for shm

The acceleration $a$ of an object that is performing simple harmonic motion is:
- always directed toward the equilibrium position **O** shown in Figure 4.20
- proportional to the displacement $x$ of the object or particle from the equilibrium position.

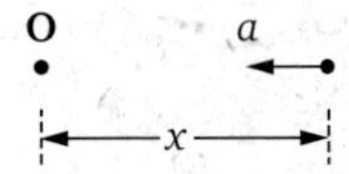

**Figure 4.20**

This condition is represented mathematically by:

$$a = -(\text{constant})x$$

The constant can be shown to be $(2\pi f)^2$ so the equation becomes:

$$a = -(2\pi f)^2 x$$

$2\pi f$ (or $\dfrac{2\pi}{T}$) is called the angular frequency and is given the symbol $\omega$. The equation is therefore often written:

$$a = -\omega^2 x$$

(To understand how this arises it is necessary to study the link between circular motion and shm. This is not a requirement for the course but is included in **Appendix A** for those who wish to know more.)

## Variation of displacement with time

When the acceleration varies with displacement according to $a = -(2\pi f)^2 x$ the variation of displacement with time is given by:

$$x = A \cos(2\pi f)t$$

The graph of displacement $x$ against time $t$ is a cosine graph if timing is assumed to commence when the displacement is at a maximum.

Figure 4.21 shows the $x$-$t$ graph for an oscillator with an amplitude of 0.1 m and a frequency of 10 Hz.

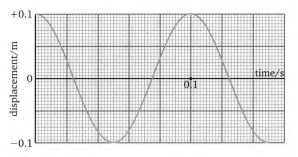

**Figure 4.21**

This variation of $x$ with $t$ can be modelled using a spreadsheet by a 'step and repeat' method. The principles involved are from your AS course. Although not required for examinations, the 'step and repeat' method is a useful technique to learn and is included in **Appendix B** for interested students. Using the model you can study oscillations with and without damping (see page 35).

> **Task**
>
> Plot on graph paper one complete cycle of the displacement–time graph for an oscillator of amplitude 0.050 m and period 2.0 s.

## Variation of velocity and acceleration with time

The relative shapes of the velocity–time and acceleration–time graphs can be deduced from the displacement–time graph. They are shown in Figure 4.22.
- The velocity of the oscillating object at any instant is the slope of the displacement–time graph.
- The acceleration at any instant is the slope of the velocity–time graph.

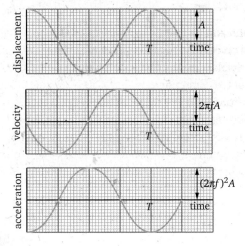

**Figure 4.22**

## Maximum speed of an oscillator

The maximum speed $v_{max}$ occurs whenever the oscillating object passes through the centre of the oscillation (the equilibrium position) and is given by:

$$v_{max} = 2\pi fA$$

where $A$ is the amplitude of the oscillator.

## Maximum acceleration of an oscillator

The maximum acceleration $a_{max}$ occurs when the displacement is at a maximum (i.e. when it is equal to the amplitude $A$ of the oscillator) and is given by:

$$a_{max} = (2\pi f)^2 A$$

## Practical investigation

There are many alternative arrangements for observing the shape of the $x$-$t$ graph.

For example, a position sensor can be attached to the object and the variation monitored using a computer.

Figure 4.23 shows one arrangement for observing the $x$-$t$ variation for a mass–spring system.

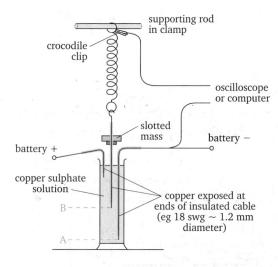

**Figure 4.23**

The system is essentially a potential divider. The potential difference between **A** and **B** is proportional to the distance between **A** and **B**.

$$V = \frac{R_{AB}}{R_{AC}}V_S = \frac{AB}{AC}V_S$$

As the mass oscillates vertically, the potential difference varies and the variation with time can be monitored using an oscilloscope or a computer.

This arrangement is ideal for investigating the effect of damping on the variation of amplitude with time for a mass–spring system (see page 35).

(see page 35)

## Test your understanding

1. Calculate the frequency of an oscillator of period $1.2 \times 10^{-3}$ s.
2. Calculate the period of a wave of frequency 35 Hz.
3. Draw a graph showing 2 cycles of the variation of displacement with time for a simple harmonic oscillator of frequency 50 Hz.
4. The equation for the displacement, in m, for a simple harmonic oscillator is:
   $$x = 1.5 \cos 30t$$
   Determine the amplitude and period of the oscillations.
5. The force acting on a mass of 0.25 kg when displaced a distance of 0.020 m from its equilibrium position is 4.9 N.
   (a) Calculate the acceleration in this position.
   (b) Determine the value of the constant $\omega$ in the general equation for shm.
   (c) Calculate the frequency of the oscillator.
6. A pendulum has a period of 2.0 s and an amplitude 0.15 m. Calculate:
   (a) the maximum speed of the pendulum bob
   (b) the maximum acceleration of the bob.

## Mass–spring System

Figure 4.24 shows a mass–spring system in which a mass is held between two stretched springs.

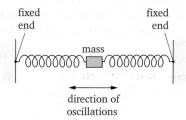

**Figure 4.24**

The period $T$ of this oscillator depends on:
- the mass $m$
- the stiffness $k$ of the spring or springs.

## Qualitative analysis

If the spring stiffness is kept constant, increasing the mass in a mass–spring:
- reduces the acceleration at a given displacement since the force for a given displacement is constant ($a = F/m$)
- reduces the average speed as the mass moves from its maximum displacement to zero
- increases the time taken to reach zero
- increases the period.

If the mass is kept constant, increasing the spring stiffness:
- increases the acceleration for a given displacement
- increases the average speed of the mass as it moves from its maximum displacement to zero
- reduces the time taken to reach zero
- reduces the period.

Since the period depends on $m$ and $k$:

$$T \propto m^p$$
$$T \propto k^q$$

where $p$ and $q$ are constants.

### Determining the constants experimentally

$p$ can be found experimentally as follows:
- obtain values of $T$ for different values of $m$ as covered in your AS course
- plot graph of log $T$ against log $m$ (this should be a straight line of positive slope)
- determine the slope of the graph
- this is $T \propto m^{\frac{1}{2}}$ (the slope should be $= \frac{1}{2}$)

Hence $T \propto m^{\frac{1}{2}}$

A similar experiment can be performed keeping the mass constant and varying the spring stiffness. The easiest way to vary the stiffness is to use a number of similar springs in series, which decreases the stiffness, or in parallel, which increases the stiffness.

(If $k$ is the stiffness of one spring:
- the stiffness for $N$ springs in series $= \frac{k}{N}$
- the stiffness of $N$ springs in parallel $= Nk$.)

The graph of log $T$ against log $k$ is a straight line of negative slope $(-\frac{1}{2})$.

Hence $T \propto k^{-\frac{1}{2}}$

The formula for the period $T$ of a mass–spring oscillator is:

$$T = 2\pi\sqrt{\frac{m}{k}}$$

### Analysing the mass–spring system

Although you will not need to be able to reproduce this analysis in an examination you should appreciate how basic physics principles are applied in the analysis.

To show that an oscillator is performing shm it is necessary to show that $a = -(\text{constant})x$.

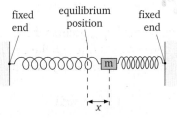

combined spring stiffness $= k$

**Figure 4.25**

At a given displacement $x$ (see Figure 4.25),

the force on the mass $F = -kx$

The minus sign shows that the direction of the force is always opposite to the direction of the displacement.

Acceleration of the mass $a = \dfrac{F}{m} = -\dfrac{kx}{m}$

The constant in the equation is therefore $\dfrac{k}{m}$.

Comparing this with the general equation for shm $(a = (2\pi f)^2 x)$ leads to:

$$\left(2\pi f\right)^2 = \frac{k}{m}$$

$$\text{so } f = \frac{1}{2\pi}\sqrt{\frac{k}{m}} \text{ and } T = 2\pi\sqrt{\frac{m}{k}}$$

## Atoms in a Lattice

Atoms in the lattice structure of a crystal are held in position by inter-atomic bonds (Figure 4.26).

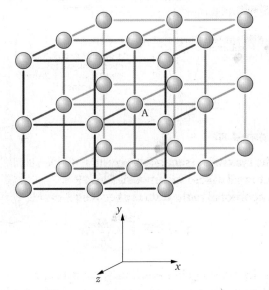

Atom A can vibrate with displacement components in directions $x$, $y$ and $z$

**Figure 4.26**

The bonds behave in a similar way to the springs in the mass–spring system. The main difference is that the atom can vibrate in three dimensions. The frequency of vibration of the atom in the lattice depends on the mass of the atom and the stiffness of the bonds. A measurement of this frequency can be made from observations of absorption spectra. The stiffness of the bonds can then be determined if the atomic mass is known.

For example, in organic substances bonds between atoms may be single or double bonds as shown in Figure 4.27.

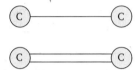

**Figure 4.27**

It is possible to determine whether the substance contains single or double bonds since the carbon atoms vibrate at a higher frequency when there are double bonds.

The double bonds behave like two springs in parallel producing twice the stiffness. This leads to a frequency that is $\sqrt{2}$ greater than for single bonds.

### Simple pendulum

A simple pendulum is one that has a heavy but small mass (a bob) on the end of a light suspension.

The period of the pendulum depends on:
- the length of the pendulum $l$
- the gravitational field strength $g$.

Note that the period does not depend on the mass of the bob.

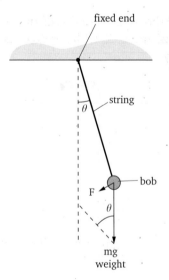

fixed end

string

$\theta$

bob

F

$\theta$

mg
weight

force toward equilibrium
position = $mg \sin \theta$

**Figure 4.28**

As shown in Figure 4.28, the force toward the equilibrium position for a given displacement is the component of the weight of the pendulum bob toward the equilibrium position.

Increasing the length:
- reduces the force at a particular displacement
- reduces the acceleration at a particular displacement
- reduces the average speed as the mass moves from its maximum displacement to zero
- increases the time taken to reach zero
- increases the period.

If the gravitational field strength is greater:
- the acceleration toward the equilibrium position increases
- the time to reach zero displacement decreases
- the period decreases.

It can therefore be predicted that:

$$T \propto l^n$$
$$T \propto g^m$$

where $n$ and $m$ are constants.

Measurements of $T$ for pendulums of different lengths $l$ can be made and a graph of log $T$ against log $l$ plotted. This should be a straight line of slope $\frac{1}{2}$.

Gravitational field strength cannot be varied in the laboratory so it is not possible to demonstrate that $T \propto 1/\sqrt{g}$.

A pendulum, however, provides an accurate method for determining $g$ in a laboratory. Geophysicists can use pendulums to map the variation of $g$ over a region that is thought to contain useful minerals. A small increase in $g$ will suggest that there is a dense body of ore containing rocks near to the Earth's surface.

### Measurement of $g$

- Obtain values of $T$ for different values of $l$ using the techniques covered in your AS course.
- Plot graph of $T^2$ against $l$ (this should be a straight line of positive slope).
- Determine the slope of the graph.

The equation for the period of a simple pendulum is:

$$T = 2\pi \sqrt{\frac{l}{g}}$$

Squaring both sides gives:

$$T^2 = 4\pi^2 \frac{l}{g}$$

Comparing this with $y = mx$ shows that the slope of the graph is:

$$\frac{4\pi^2}{g}$$

## Test your understanding

1 Calculate the period of a simple pendulum that has a length of 0.50 m:
   (a) on Earth
   (b) on the Moon, where the acceleration due to gravity is one sixth that on Earth.

2 A pendulum has a period of 1.50 s. Determine the period when:
   (a) the length is doubled
   (b) the mass of the bob is halved.

3 (a) Calculate the length of the pendulum in question 2.
   (b) What would be the length if the period were increased to 3.0 s?

4 The period of oscillation of a mass of 0.60 kg in a mass–spring system is 2.5 s. Calculate the stiffness of the spring used in the system.

5 An atom in a lattice has a mass of $5.0 \times 10^{-26}$ kg. When a beam of electromagnetic radiation passes through the lattice, the infra red wavelength at $6.0 \times 10^{-5}$ m causes this atom to resonate so that energy is absorbed from the beam. Determine the stiffness of the bonds that hold the atom in place. ($c = 3 \times 10^8$ m s$^{-1}$)

## Phase Difference

Oscillators that reach their maximum displacement in the same direction at the same time are said to be oscillating **in phase**. When they reach the maximum displacement at different times they are said to be **oscillating out of phase**.

Two oscillators in which one reaches a maximum displacement in one direction at the same time as the other reaches a maximum in the other direction are said to be **antiphase** (or that their phase difference is $\pi$ radians, 180°). The phase difference of other oscillators is defined by the angle by which one reaches its maximum before the other. Figure 4.29 shows the x-t graphs for oscillators that are oscillating in phase, in antiphase, and with a phase difference of $\frac{\pi}{2}$ radians or 90°.

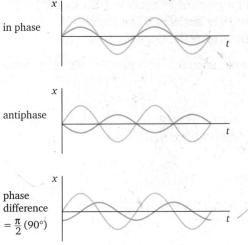

**Figure 4.29**

Oscillators of the same frequency always retain the same phase difference. If they have different frequencies they move in phase, then out of phase, and then back in phase again.

## Energy of an Oscillator

Unless there is a loss of energy to the surroundings (see damping on page 34) the **total energy** of an oscillator remains constant.

When the oscillator is at its maximum displacement (A), the energy is all **potential energy $E_p$**.

At the centre of the oscillation, the position of zero displacement, the energy is all **kinetic energy $E_k$**.

Between these two positions, the energy is continually being transferred from PE to KE, and back again to PE. The total energy is the sum of the potential and kinetic energy.

Since the maximum speed is $2\pi fA$, the maximum kinetic energy is:

$$\tfrac{1}{2}m(2\pi fA)^2$$

For a mass–spring system, the maximum potential energy will be when the springs are stretched or compressed the most. The maximum potential energy is:

$$\tfrac{1}{2}kA^2$$

$$\text{(using } E = \tfrac{1}{2}k(\Delta l)^2)$$

For a given oscillator $m$, $k$ and $f$ are constant so the total energy of the oscillator is proportional to the square of the amplitude, $A^2$.

This is true for all sinusoidal oscillators.

### Variation of $E_p$ and $E_k$

Figure 4.30 shows the variation of kinetic energy $E_k$ and potential energy $E_p$ with position $x$. $E_T$ is the total energy.

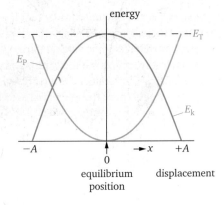

**Figure 4.30**

Figure 4.31 shows the variation of $E_k$ and $E_p$ with time $t$.

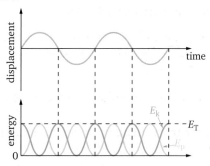

**Figure 4.31**

The corresponding displacement–time graph is also shown. At $t = 0$, the oscillator is at zero displacement.

### Test your understanding

1 $A$ is the amplitude of an oscillator. Show that:
   (a) when the displacement is $\frac{A}{2}, \frac{1}{4}$ of the total energy is potential energy
   (b) $E_k = E_p$ when the displacement of the oscillating object is $0.71A$.
2 A pendulum bob has a mass of 0.25 kg. The period of the pendulum is 2.5 s.
   (a) Calculate the length of the pendulum.
   (b) The amplitude of the oscillations is 0.20 m. Calculate the total energy of the oscillator.
   (c) Calculate the total energy when the amplitude falls to 30% of the initial amplitude.
3 When an oscillator has an amplitude of 0.15 m its total energy is 2.0 J. Calculate the amplitude when its energy has fallen to 0.50 J.
4 Calculate the maximum kinetic energy of an oscillating 1.5-kg mass that has a frequency of 2.5 Hz and an amplitude of 0.012 m.
5 (a) A body performing simple harmonic motion undergoes a cycle of accelerations. State the conditions relating to the acceleration that are essential for simple harmonic motion.
   (b) When displaced, a system such as that in Figure 4.32 undergoes simple harmonic motion provided that the vertical force exerted by the springs is directly proportional to the vertical displacement. State briefly how you would check the range over which this condition holds for the system.

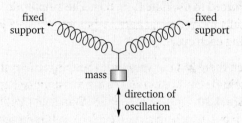

**Figure 4.32**

(c) It is thought that a spider can tell when a fly is trapped in its web by the change in the frequency of vibration of the web. One web oscillates at a frequency of 15 Hz when there is no fly on it. The effective mass of the web is $1.2 \times 10^{-4}$ kg.
   (i) Calculate the effective stiffness $k$ of the web assuming that it behaves like the mass–spring system in Figure 4.32.
   (ii) Determine the new frequency that the spider detects when a fly of mass $5.0 \times 10^{-5}$ kg is trapped in the centre of the web.
   (iii) The amplitude of the oscillations is 7.5 mm. Calculate the maximum speed of the fly during the oscillation.
   (iv) Calculate the total energy of the oscillations.

## Damping

In practical oscillators, energy is lost to the surroundings as the body oscillates.

Damping may be caused by:
- friction when a solid body oscillates while in contact with another solid body
- viscous forces when a solid body is in contact with a fluid (liquid or gas).

The result is that the total energy of the oscillator, and hence the amplitude of the oscillations, decreases with time.

### Degree of damping

An oscillator that loses a large proportion of its total energy per oscillation is **heavily damped**. The amplitude of oscillations of a heavily damped oscillator falls quickly.

An oscillator that loses a small proportion of its total energy per oscillation is **lightly damped**.

The period of oscillation of a lightly damped oscillator is the same as that of the oscillator with no damping. The period remains constant as the amplitude falls.

EXAMPLE: The amplitude of an oscillator falls to 0.80 of its original amplitude after one complete oscillation.

Calculate the proportion of the original energy that has been transferred to the surroundings.

Energy is proportional to $A^2$ so new energy is $0.8^2$ of the original energy.

Energy lost = 0.36 of the initial energy

A **critically damped** oscillator is one in which the energy is transferred to the surroundings very rapidly. The oscillator either does not actually oscillate at all or only just overshoots the equilibrium position before coming to rest at the equilibrium position. This type of damping is built in to moving coil meters so that they show a stable new reading quickly.

It is also a design feature of a car shock absorber so that when a car goes over a bump on the road the oscillation dies away immediately.

Figure 4.33 shows how the displacement varies with time for an oscillator that is heavily damped, lightly damped and critically damped.

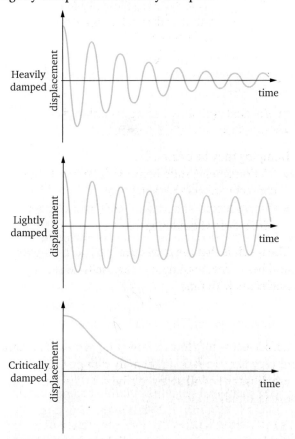

**Figure 4.33**

Investigation of damping

A suitable apparatus is shown in Figure 4.23. Different degrees of damping can be investigated by varying the resistance to the motion and monitoring the change in amplitude using a computer.

## Forced Oscillations

Forced oscillations occur when a system is made to oscillate by a periodic external force. Energy is then supplied to the oscillator at regular intervals. The energy may be supplied:
- by a force that produces short impulses such as when pushing a swing
- by a force that varies sinusoidally.

The external source of energy is referred to as the **driver oscillator**. The body that is made to oscillate is referred to as the **driven oscillator**.

When the force due to the driver varies sinusoidally, the driven oscillator oscillates at the same frequency as the driver.

The amplitude of the oscillations produced depends on:
- how near the driver frequency is to the natural frequency of the driven oscillator
- the degree of damping present in the driven oscillator.

Figure 4.34 shows how the amplitude varies with driver frequency for a heavily and lightly damped driven oscillator.

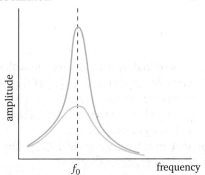

**Figure 4.34**

## Resonance

The amplitude of the driven oscillations increases as the frequency approaches the natural frequency of the driven oscillator.

When the driver and driven oscillator frequencies are equal (shown as $f_o$ on Figure 4.34) the amplitude is a maximum. This is the condition for **resonance**.

Notice that:
- For a given driver frequency the amplitude is lower if there is heavier damping
- the resonance is sharper for lighter damping
- the resonant frequency is not affected significantly by damping (for very heavy damping, the resonant frequency decreases).

Figure 4.35 shows how oscillations build up when resonant systems with different degrees of damping are forced to oscillate from rest. The amplitude initially increases as energy is put into the oscillator during each cycle.

When there is no or very little damping, the system amplitude continues to increase as shown in Figure 4.35(a). In a lightly damped mechanical structure, (such as a poorly designed bridge), this increase could lead to structural damage.

Figure 4.35(c) is the most heavily damped of the three systems. The figures show that more damping lowers the final amplitude of the oscillations.

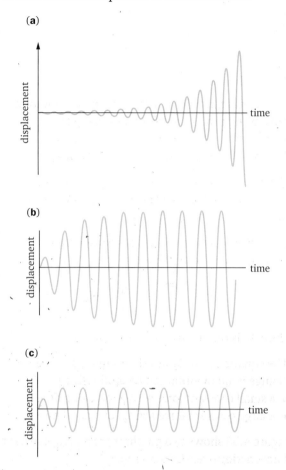

**Figure 4.35**

The amplitude stops increasing when the energy input from the driver in one cycle is equal to that lost to the surroundings from the driven oscillator during one cycle.

### Investigating forced oscillations

Figure 4.36 shows an apparatus for such an investigation. A vibrator could also be incorporated into the apparatus in Figure 4.23 to monitor the oscillations and record the change of amplitude as the oscillations build up from rest.

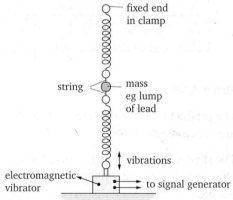

**Figure 4.36**

The mass–spring system should have a natural frequency of at least 7 Hz (a 0.2 kg mass between two expendable springs is suitable). The vibrator is driven by a signal generator. This should be capable of producing oscillations as low as 1 Hz.

The output voltage of the oscillator is kept constant and the maximum amplitude of the oscillations is measured for different frequency settings. A graph of maximum amplitude against frequency is then plotted.

For some frequencies, especially when the driver and driven frequencies become closer together, the amplitude of the oscillations builds up and then dies away before building up again. You do not need to know about these transient oscillations for the A2 examination.

The measurement you need to make is the maximum amplitude reached during the cycle of changes.

### Forced oscillations in practice

- In musical instruments such as pianos, strings other than the one struck will be forced to oscillate. The instrument case itself may also oscillate. These oscillations affect the tonal quality of the sound produced.
- Atoms trapped in a lattice may be forced to vibrate when radiation falls on them. Radiation at the natural frequency of the atom in the lattice will be absorbed by the atom that resonates. The intensity of the frequency originally present in the incident radiation will be reduced.
- Moving parts of machinery and other structures such as bridges are often forced to vibrate and, in extreme cases, will resonate. The continual vibrations and the large build-up of amplitude may cause failure of metals in the structure.

### Task

Research other instances in which forced vibrations and resonance occur in practical situations. Explain in each case whether resonance is an essential part of the design or a disadvantage in the system.

# Momentum Concepts

## Definition of Linear Momentum

Momentum is a measure of a body's inertia. Linear momentum $p$ is defined as the product of mass $m$ and velocity $v$:

$$\text{Momentum, } p = mv$$

The unit of momentum is kg m s$^{-1}$.
(For reasons that follow, an alternative unit is N s.)

Momentum is a vector quantity so it has a magnitude (size) and a direction.

Figure 4.37 shows two objects of mass 2.0 kg moving in opposite directions at the same speed, 4.0 m s$^{-1}$.

**Figure 4.37**

The magnitude of the momentum of each object is identical, 8.0 kg m s$^{-1}$ but, because they are moving in opposite directions, the difference between them is 16 kg m s$^{-1}$.

A small object moving at a high speed may have the same momentum as a large object moving at a low speed.

For example, a bullet of mass 0.14 kg moving at 1000 m s$^{-1}$ has a momentum of 140 kg m s$^{-1}$.

A person of mass 70 kg has the same momentum when travelling at $140/70 = 2.0$ m s$^{-1}$.

## Changing Momentum and Impulse

A body retains the same momentum unless a force acts on it. This is one way of stating Newton's first law of motion.

The momentum of an object can be changed by applying a force. The change in momentum $\Delta(mv)$ in kg m s$^{-1}$ is equal to the product of:
- the magnitude of the force $F$ in N
- the time for which the force acts $t$ in s.

$$Ft = \Delta(mv)$$

The change in momentum takes place in the direction of the applied force. The product $Ft$ is called an **impulse**.

Notice that the units on both sides of the equation must be equivalent so the unit of change in momentum, and therefore of momentum itself. The unit may therefore be given as N s or kg m s$^{-1}$.

The equation $Ft = \Delta(mv)$ is an alternative, and often more useful, way of stating Newton's second law $F = ma$.

For a situation in which mass is constant the equations are shown to be equivalent as follows:

$$F = \frac{\Delta(mv)}{t} = m\frac{\Delta v}{t} = ma$$

since $\Delta v$ is the change in velocity in time $t$.

The equation $Ft = \Delta(mv)$ shows that the same change in momentum can be produced by:
- a small constant force acting for a long time
- a large constant force acting for a short time.

Figure 4.38 shows two graphs representing constant forces acting over different times.

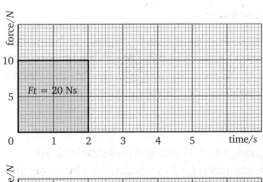

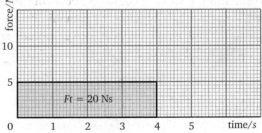

**Figure 4.38**

The product $Ft$ = momentum change. This is the area under each graph.

The shaded areas in Figure 4.38 are equal so that although the forces acting are different, the momentum change produced in each case will be identical.

EXAMPLE: An object has zero momentum.

(a) Calculate the constant force necessary to increase the momentum of the object by 15 N s in:

  (i)   2.0 s

  (ii)  0.05 s.

(b) How long would it take a constant force of 3.0 N to increase the momentum to 15 N s?

(a) (i)   $F \times 2 = 15 \therefore F = 7.5$ N

   (ii)  $F \times 0.05 = 15 \therefore F \times 300$ N

(b) $3.0 \times t = 15$
    $t = 5.0$ s

Notice that these examples could have been solved using equations of motion and $F = ma$ since the force is constant.

In many practical situations, the force producing the change in speed and hence in momentum is not constant. In this case, the equations of motion cannot easily be applied.

## Impulses for varying force

Forces in practice do not change from zero to a maximum in zero time. Figure 4.39 shows a more realistic variation of force with time during a collision between two objects.

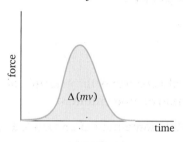

**Figure 4.39**

Such variations occur when a golf ball is struck or when a vehicle crashes into a wall.

As in Figure 4.38, the area under the curve still represents the momentum change. This can be determined by:
- counting the number of squares under the curve
- calculating the momentum change represented by one square
- multiplying these together.

Knowing the mass involved, the change in velocity can be determined despite the fact that the acceleration varies.

EXAMPLE: Figure 4.40 shows the force–time graph when a ball of mass 0.35 kg is kicked.

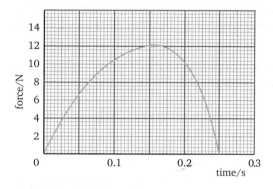

**Figure 4.40**

Calculate:
(a) the momentum change of the ball:

(b) its speed immediately after being kicked

(c) the average acceleration of the ball.

(a) Approximate number of large squares under the graph = 43.
Momentum change per square = 0.050 N s.
Total momentum change = 2.5 N s.

(b) $mv = 2.5$
   $0.35\, v = 2.5$
   $v = 7.1\ \mathrm{m\,s^{-1}}$

(c) $a = \dfrac{\Delta v}{\Delta t} = \dfrac{71}{0.25} = 29\,\mathrm{m\,s^{-2}}$

## Rebounding Objects

Consider the case of two bodies **A** and **B** of equal mass $m$ hitting the ground with the same velocity $v$ as shown in Figure 4.41.

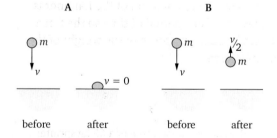

**Figure 4.41**

**A** sticks to the ground and comes to rest.

**B** rebounds with half its original speed but in the opposite direction.
Which exerts the greater impulse?

Change in momentum of **A** $= mv$

Change in momentum of **B** $= mv + m\dfrac{v}{2} = 1.5mv$

So **B** exerts the greater impulse. If the time to bring **A** to rest is similar to the time for which **B** is in contact with the ground then the force exerted on the ground by **B** is greater than that exerted by **A**.

The calculation of change in momentum of rebounding particles is used in deriving formulae using the kinetic theory of gases (see page 47).

## Practical Applications

### Crumple zones

Cars are designed so that the vehicle is brought to rest over a long period by allowing parts of the car to collapse when a crash occurs. The force experienced by passengers is therefore reduced.

This is similar to the protection of eggs in egg boxes. When a dropped box hits the floor, cardboard sections in the box collapse. This reduces the force experienced by the eggs.

### Seat belts and air bags

These serve a similar purpose. Instead of the momentum falling to zero in a short time, the duration is extended so that the force experienced is reduced. With seat belts, there is inelastic deformation of the belt material. Using air bags, the gas compresses over time to reduce the force.

### Running shoes

The soles of running shoes deform when the foot impacts with a solid surface. This reduces the force on the leg joints reducing the risk of injury.

### Using hammers

In this case, the momentum of the hammer is reduced in a short period of time so the force exerted is much greater than the weight of the hammer head.

### Test your understanding

1  Calculate the momentum of a 65 kg sprinter when travelling at 9.5 m s$^{-1}$.
2  Calculate the velocity of a car of mass 700 kg that has the same momentum as the sprinter in Question **1**.
3  A force of 2000 N is applied for 3.5 s to a spacecraft of mass 12 000 kg.
   (a) Calculate the impulse the spacecraft receives.
   (b) Calculate the change in velocity of the spacecraft.

4  Explain in terms of the physics involved why a sand pit is used to protect athletes when landing in the high jump.
5  An egg box is designed to protect eggs of total mass 0.36 kg when landing on a hard surface at 5 m s$^{-1}$.
   The impact time is 0.15 s. Calculate the average force acting during the impact.
6  A pool ball of mass 0.15 kg hits a cushion at a speed of 3.5 m s$^{-1}$ and rebounds at a speed of 3.0 m s$^{-1}$. Calculate the change in momentum of the pool ball during the impact.
7  Figure 4.42 shows how the force acting on a tennis ball of mass 0.060 kg varies with time when it is struck.

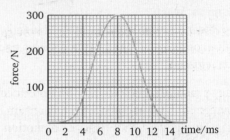

**Figure 4.42**

Estimate:
   (a) the total impulse given to the ball
   (b) the velocity of the ball on leaving the racquet
   (c) the constant force that would produce the same change in momentum in the same impact time.

## Conservation of Momentum

The general statement of the **principle of conservation of momentum** is:

> **the total momentum of an isolated system is constant.**

An isolated system is one on which no external forces are acting. Within such a system, momentum can be exchanged between the bodies but overall there is no change.

### Collisions

Momentum is conserved for **all** types of collision.

In this course you need only consider head-on collisions (or interactions) between two bodies for which the momenta are along the same straight line in either the same direction or in opposite directions. Such collisions occur, for example, between vehicles on a linear air track as shown in Figure 4.43.

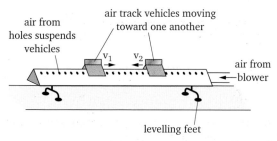

**Figure 4.43**

### 'Explosions'

You also need to apply the principle to situations in which two masses fly apart having been provided with energy from within the system. These occur, for example, when:

- two linear air track vehicles are held together with a compressed spring between them and are then released
- a radioactive particle undergoes decay by alpha emission (Figure 4.44)
- a rocket is propelled by ejecting high velocity gases (Figure 4.45).

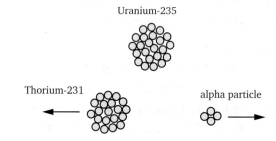

**Figure 4.44**

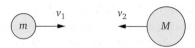

**Figure 4.45**

### Applying the principle

Figure 4.46 shows the velocities $v_1$ and $v_2$ of two objects of masses $m$ and $M$ before they collide. They are initially moving toward each other. Figure 4.47 shows their velocities $v_3$ and $v_4$ after the collision. It is assumed that they will both move to the right.

**Figure 4.46**

**Figure 4.47**

Taking a momentum to the right as positive, the total momentum before the collision
$= mv_1 - Mv_2$, and the total momentum after the collision $= mv_3 + Mv_4$.

Using the principle of conservation of momentum:

$$mv_1 - Mv_2 = mv_3 + Mv_4$$

Note that a negative value for a velocity or momentum would mean the object is moving to the left.

EXAMPLE: Figure 4.48 shows the masses and initial velocities of two bodies about to collide.

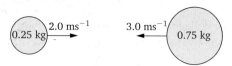

**Figure 4.48**

After the collision the 0.25 kg mass moves at $4.0 \text{ m s}^{-1}$ to the left. Calculate the velocity of the 0.75 kg mass.

Total momentum before the collision
$$= 0.25 \times 2.0 - 0.75 \times 3.0 = -1.75 \text{ N s}$$
(i.e. the total momentum is 1.75 N s to the left)

Total momentum after the collision
$$= -0.25 \times 4.0 + 0.75v$$

Using momentum conservation:
$$\begin{aligned} -1.0 + 0.75v &= -1.75 \\ 0.75v &= -1.75 + 1.0 \\ &= -0.75 \\ v &= -1.0 \text{ m s}^{-1} \end{aligned}$$

The final velocity of the 0.75 kg mass is $1.0 \text{ m s}^{-1}$ to the left.

### Momentum conservation and Newton's laws

The principle of conservation of momentum is a necessary result of applying Newton's laws to the collision.

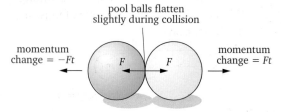

**Figure 4.49**

Figure 4.49 shows a collision between two pool balls. During this collision:

- at all times each ball experiences a force of the same magnitude $F$ but in opposite directions (From Newton's third law: action and reaction are equal and opposite)
- the forces must act for the same time $t$ on each ball
- each ball experiences an equal but opposite impulse $Ft$
- each ball has the same change in momentum but in opposite directions [from Newton's second law: $Ft = \Delta(mv)$].

It follows that, since the change in momentum of each ball is equal and opposite, there is no change in the total momentum of the two balls.

## Elastic and Inelastic Collisions

The **total energy** of a system remains the same for **all** collisions. In an **elastic collision** there is no change in total **kinetic** energy (KE) of the system of bodies.

All other collisions are **inelastic**. In inelastic collisions energy is transferred into forms other than KE. Therefore the total KE falls.

The 'lost' KE may be used to deform the bodies permanently (as in a collision between two cars) and/or raise their temperature by increasing their internal energy.

Most practical collisions are inelastic. Only collisions between atoms or between subatomic particles are truly elastic.

### Task

Use the data and answer to the example on page 40 to show that the collision is inelastic.

## Summary

In calculations remember the following:
- In **elastic collisions**, momentum, kinetic energy and total energy are conserved;
- In **inelastic collisions**, momentum and total energy are conserved.

## Some Special Cases

Figure 4.50 shows a particle of mass $m_1$ colliding with a velocity $v$ with another particle of mass $m_2$ that is initially stationary.

initial velocity = 0

**Figure 4.50**

### Case 1

When $m_1 = m_2$ and the collision is elastic, after the collision, $m_1$ is stationary and $m_2$ moves with velocity $v$.

This is the only way that both KE and momentum can be conserved. All the momentum and KE of $m_1$ has been passed on to $m_2$.

### Case 2

When $m_1$ and $m_2$ in Figure 4.50 stick together when they collide so that they move together at the same velocity after the collision:

$$\text{velocity after the collision} = \frac{m_1}{(m_1 + m_2)}.v$$

When $m_1 = m_2$, the velocity after the collison is $\frac{v}{2}$.

### Case 3

A single stationary mass $M$ explodes into two parts. The mass of one of the parts is $m$.
- The momentum of the two parts before the explosion = 0.
- The two parts move in opposite directions after the explosion.
- The magnitude of the momentum of each part is the same.
- $mv_1 = (M-m)v_2$, where $v_1$ is the speed of $m$ after the explosion and $v_2$ the speed of the other mass.

EXAMPLE: The mass of the recoil nucleus that remains after a stationary radioactive particle has emitted an alpha particle of mass $6.8 \times 10^{-27}$ kg is $3.9 \times 10^{-25}$ kg. The alpha particle has an energy of $4.9 \times 10^{-16}$ J. Calculate:

(a)   the speed of the alpha particle

(b)   the speed of the recoil nucleus.

(a)   Using $E_K = \frac{1}{2}mv^2$

$4.9 \times 10^{-16} = \frac{1}{2} \, 6.8 \times 10^{-27} \, v^2$

$v = 3.8 \times 10^5$ m s$^{-1}$

(b)   Using momentum conservation

$6.8 \times 10^{-27} \times 3.8 \times 10^5 = 3.9 \times 10^{-25} \, v$

speed of recoil nucleus $= 6.6 \times 10^3$ m s$^{-1}$

### Test your understanding

1   A 0.050 kg lump of Plasticine moving at 15 m s$^{-1}$ sticks to the 0.25 kg bob of a stationary pendulum. Calculate the speed of the combined mass immediately after the collision.

2   A 0.40 kg air track vehicle **A** travelling at 3.0 m s$^{-1}$ collides with a 0.80 kg stationary vehicle **B**. After the collision the 0.040 kg vehicle moves in the original direction at a speed of 0.5 m s$^{-1}$.
    (a)  Calculate the velocity of **B** after the collision.
    (b)  Show that the collision is inelastic.

3   Two perfectly elastic balls of equal mass move toward each other at equal speeds and collide head on. State and explain what happens following the collision.

4   In an explosion the exploding object breaks up into two parts **A** and **B**. **A** has three times the mass of **B**. The speed of **A** after the explosion is 20 m s$^{-1}$.
    (a)  Calculate the speed of **B**
    (b)  Calculate the percentage of the total energy carried by **A** as KE after the explosion.

# Gas Laws

## State of a Gas

The physical condition or **state** of a gas is described by its:

- pressure $p$
- volume $V$
- temperature $T$
- mass (which is proportional to the number of molecules it contains).

When describing the state:

- the pressure is in Pa where $1 \, \text{Pa} = 1 \, \text{N m}^{-2}$
- the volume is in $\text{m}^3$
- the temperature is in kelvin K.

$$T/\text{K} = T/°\text{C} + 273.15$$

In A2, the use of three significant figures (273) is sufficient when converting units.

## Gas Laws

The gas laws are the relationships between the pressure $p$, volume $V$ and temperature $T$ for a fixed mass of a gas.

(Note that it is the relationships that are important. You do not need to remember the names of the scientist associated with each law.)

### $p \propto \dfrac{1}{V}$ when $T$ is constant

This is Boyle's law which, in words, states that the pressure of a fixed mass of gas is inversely proportional to its volume when the temperature is kept constant.

Figure 4.51 shows a graph of $p$ against $V$ for a temperature $T_1$ and that for a higher temperature $T_2$.

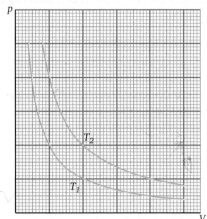

**Figure 4.51**

In a practical investigation to test the relationship, $p$ is plotted against $\dfrac{1}{V}$ as shown in Figure 4.52.

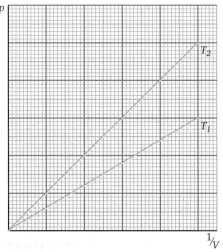

**Figure 4.52**

The two lines shown are for a temperature $T_1$ and a higher temperature $T_2$ for the same gas. For any given volume, the pressure is higher when the temperature is higher.

Boyle's law is usually quoted and used in the forms:

$$pV = constant$$

or

$$p_1 V_1 = p_2 V_2$$

where $p_1$ and $V_1$ are the values of pressure and volume before the change and $p_2$ and $V_2$ are the values after the change.

Boyle's law is obeyed at constant temperature. This is called an **isothermal change**.

For a change to be isothermal any change in internal energy that occurs as a result of a change in volume has to be lost to, or gained from, the surroundings. For this to occur:

- time has to be allowed for the temperature to return to the original value
- the container of the gas should be a good conductor of heat.

### $V \propto T$ when p is constant

This is Charles' law which, in words, states that the volume of a fixed mass of gas is proportional to the temperature in K when the pressure is kept constant.

Charles' law is often written as:

$$\frac{V}{T} = constant$$

or

$$\frac{V_1}{T_1} = \frac{V_2}{T_2}$$

where $V_1$ and $T_1$ are the values of volume and temperature before the change and $V_2$ and $T_2$ are the values after the change.

Figure 4.53 shows a graph of $V$ against $T/K$.

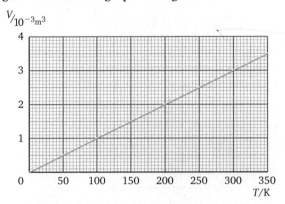

**Figure 4.53**

Figure 4.54 shows the graph when the temperature is in °C.

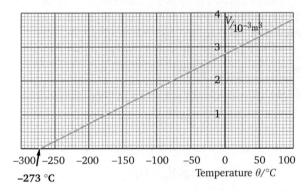

**Figure 4.54**

In Figure 4.54 the intercept on the temperature axis is at absolute zero or −273°C. This point is zero on the Kelvin temperature scale.

## $p \propto T$ when $V$ is constant

This is Gay-Lussac's law and is similar to Charles' law. It relates pressure to the temperature in K for constant volume.

$$\frac{p}{T} = constant$$

or

$$\frac{p_1}{T_1} = \frac{p_2}{T_2}$$

The graphs in this case are similar to the $V$-$T$ graphs shown in Figures 4.53 and 4.54 replacing $V$ with $p$ on the $y$ axis.

The gas laws are summarised for a fixed mass of gas by the equations:

$$\frac{pV}{T} = constant \text{ or } \frac{p_1V_1}{T_1} = \frac{p_2V_2}{T_2}$$

where $p_1$, $V_1$ and $T_1$ are the values of pressure, volume and temperature (in K) before the change and $p_2$, $V_2$ and $T_2$ are the values after the change.

EXAMPLE: Air in a bicycle pump has a pressure of $1.00 \times 10^5$ Pa and temperature 295K. The air is compressed from its initial volume of $4.00 \times 10^{-4}$ m³ to a volume of $2.50 \times 10^{-4}$ m³ while a finger is held over the pump outlet.

(a) Calculate the new pressure when the temperature is kept constant.

(b) Calculate the new pressure if the temperature rises to 303K.

(a)
$$p_1V_1 = p_2V_2$$
$$1.0 \times 10^5 \times 4.0 \times 10^{-4} = p_2 \times 2.5 \times 10^{-4}$$
$$p_2 = 1.60 \times 10^5 \text{ Pa}$$

(b)
$$\frac{p_1V_1}{T_1} = \frac{p_2V_2}{T_2}$$
$$\frac{\left(1.0 \times 10^5\right) \times \left(4.0 \times 10^{-4}\right)}{295} = \frac{p_2 \times \left(2.5 \times 10^{-4}\right)}{303}$$
$$p_2 = 1.64 \times 10^5 \text{ Pa}$$

## Pressure

Pressure $p$ is defined as force per unit area (i.e. force $F$ divided by the area $A$ on which the force acts).

$$p = \frac{F}{A}$$

The unit of pressure is the pascal Pa.

$$1 \text{ Pa} = 1\text{N m}^{-2}$$

Normal atmospheric pressure is $1.01 \times 10^5$ Pa.

Pressures are often given in 'metres of mercury' as measured by the height $h$ of mercury in a mercury barometer (see Figure 4.55). This is converted to Pa by the equation:

$$p = h\rho g$$

where $\rho$ is the density of mercury in kg m$^{-3}$

$g$ is the gravitational field strength. (9.8 N kg$^{-1}$)

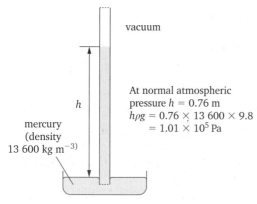

**Figure 4.55**

The pressure $p$ at a given depth $h$ in a liquid of density $\rho$, with its upper surface exposed to atmospheric pressure $p_A$, as shown in Figure 4.56, is given by:

$$p = p_A + h\rho g$$

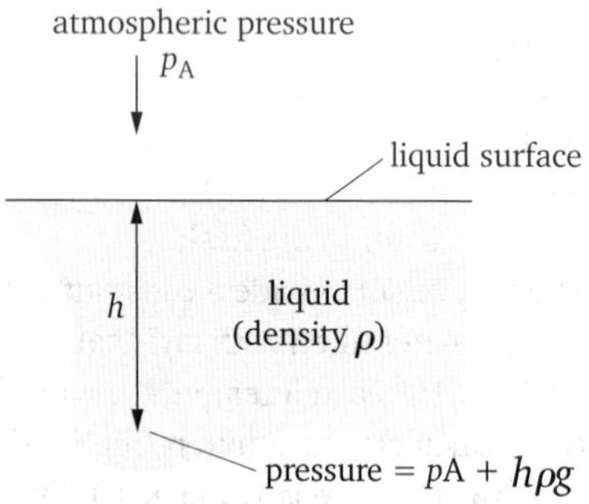

atmospheric pressure
$\downarrow p_A$

liquid surface

$h$   liquid (density $\rho$)

pressure $= p_A + h\rho g$

**Figure 4.56**

## The Ideal Gas Equation

- For any ideal gas that contains $n$ mol of the gas, the value of $\dfrac{pV}{T}$ is always the same.
- The value is $nR$ where $R$ is the **molar gas constant**.
- The ideal gas equation relates the pressure, volume, temperature and amount of gas:
$$\frac{pV}{T} = nR$$
or
$$pV = nRT$$
- When pressure is in Pa, volume in m³ and temperature in K, $R = 8.3 \, \text{J mol}^{-1}\text{K}^{-1}$.

**Remember that you must use SI units in calculations.**

**Real gases** behave like ideal gases at low pressures. For real gases at high pressures, the equation has to be modified. In this course, all gases are assumed to be ideal gases even when the low pressure condition is not satisfied.

### Molar mass and the Avogadro constant

**1 mole** (1 mol) is defined as the number of atoms in 12 g of carbon-12.

This number is called the **Avogadro constant $N_A$** and is $6.02 \times 10^{23} \, \text{mol}^{-1}$.

The molar mass of any substance is the mass in g that contains $6.02 \times 10^{23}$ atoms or molecules.

To a degree of accuracy suitable for A2 level, the molar mass of a substance, in g, is numerically equal to the number of nucleons in the atom or molecule.

There is 1 mol ($6.02 \times 10^{23}$) atoms in:
- 1 g of $^1_1\text{H}$ (hydrogen-1),
- 4 g of $^4_2\text{He}$ (helium-4), etc.

Putting it another way:
- the molar mass of hydrogen-1 is 1g
- the molar mass of helium-4 is 4g, etc.

In questions the molar mass will be given when required. You should be able to convert mass of gas to number of moles and vice versa.

In work on gases you will need to determine the number of moles in a gas. In the work on radioactivity in Module 5 you will need to be able to determine the number of atoms in a given mass of a radioactive nuclide.

**EXAMPLE ONE:** The relative molecular mass of nitrogen is 28.

Calculate:
(a) the mass of 0.20 mol of nitrogen
(b) the number of moles in 1.2 g of nitrogen
(c) the number of atoms in 1.2 g of nitrogen.

(a) 1 mol of nitrogen has a mass of 28 g
0.2 mol has a mass of $0.2 \times 28 = 5.6$ g

(b) 28 g of nitrogen $\equiv$ 1 mol
$$1.2\text{g of nitrogen} \equiv \frac{1.2}{28} \equiv 0.043 \text{ mol}$$

(c) 1 mol contains $6.02 \times 10^{23}$ molecules
0.043 g contains $6.02 \times 10^{23} \times 0.043$
$= 2.6 \times 10^{22}$ molecules

**EXAMPLE TWO:** A gas contains 0.20 g of helium-4. The volume of the gas is 0.015 m³ and the temperature is 300 K.

Calculate the pressure of the gas.

$$0.20\text{g of helium-4} \equiv \frac{0.20}{4} \equiv 0.050 \text{ mol}$$

$$pV = nRT \text{ so } p = \frac{nRT}{V}$$

$$p = \frac{0.050 \times 8.3 \times 300}{0.015} = 8300 \text{ Pa (8.3 kPa)}$$

## Test your understanding

1  A gas occupies a volume of $2.2 \times 10^{-3}$ m³ and has a pressure of $1.2 \times 10^5$ Pa. It is compressed isothermally to a volume of $1.5 \times 10^{-3}$ m³. Calculate the new pressure of the gas.

2  The temperature of a gas rises from 290 K to 360 K at constant volume. The initial pressure was $1.01 \times 10^5$ Pa. Calculate the new pressure.

3  Complete the following table that shows pressures, volumes and temperatures before and after a process that changes the state of the gas.

**Table 4.1**

| | | | | |
|---|---|---|---|---|
| **b e f o r e** | $p/10^5$ Pa | 2.0 | 3.5 | 2.6 |
| | $V/$m³ | 0.15 | 0.0030 | 0.30 |
| | $T/$K | 315 | 300 | |
| **a f t e r** | $p/10^5$ Pa | 2.0 | | 5.5 |
| | $V/$m³ | | 0.015 | 0.40 |
| | $T/$K | 130 | 300 | 310 |

4  A gas tank has a volume of 1.5 m³ and is filled with air to a pressure of 20 times atmospheric pressure when at a temperature of 7.0°C. Calculate the volume of the air when allowed to expand to atmospheric pressure at room temperature of 27°C.

5  Figure 4.57 shows the pressure-volume graph for a fixed mass of a gas at a temperature of 300 K. $R = 8.3$ J mol⁻¹ K⁻¹.

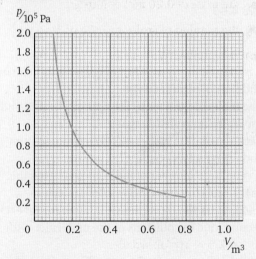

**Figure 4.57**

(a) Use measurements from the graph to show that the change takes place at constant temperature.

(b) Calculate the number of moles of gas in the sample.

(c) Sketch the isothermal for the same mass of gas at a temperature of 350 K.

6  Complete the following table to give corresponding values of molar mass, mass $m$, number of moles $n$ and number of molecules $N$. ($N_A = 6.0 \times 10^{23}$ mol⁻¹.)

**Table 4.2**

| molar mass/g | 1 | 32 | | |
|---|---|---|---|---|
| $m/$g | | | 0.35 | 3.7 |
| $n/$mol | 0.12 | | $5.8 \times 10^{-3}$ | |
| $N$ | | $3.0 \times 10^{20}$ | | $5.0 \times 10^{22}$ |

7  A gas has a pressure of $0.75 \times 10^5$ Pa and temperature 37°C. It is contained in a cylinder of volume 0.12 m³. The molar mass of the gas is 2.0 g. Calculate:
   (i)  the number of moles of gas in the container
   (ii) the mass of gas in the container.

8  Calculate the temperature of 10 mol of gas that has a pressure of $1.0 \times 10^5$ Pa when it is in a container of volume 1 m³.

9  The data on a vacuum pump suggest that the minimum attainable pressure is 150 Pa at 300 K.
   The Avogadro constant $N_A = 6.0 \times 10^{23}$ mol⁻¹
   Molar gas constant $R \quad = 8.3$ J mol⁻¹ K⁻¹
   Molar mass of air $\quad = 0.025$ kg mol⁻¹
   For a $5.0 \times 10^{-4}$ m⁻³ flask, calculate:
   (a) the number of moles of air in the flask
   (b) the number of molecules in the flask
   (c) the mass of air in the flask.

10 (a) The air pressure inside a constant volume gas thermometer filled with an ideal gas is $1.010 \times 10^5$ Pa at a temperature of 0°C.
      (i)  Calculate the temperature in °C when the pressure is $1.600 \times 10^5$ Pa.
      (ii) Calculate the pressure when the temperature is 37°C.
   (b) Figure 4.58 shows the variations of pressure with temperature for two constant volume gas thermometers **A** and **B** that contain the same volume of gas.

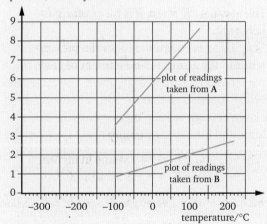

**Figure 4.58**

   (i)  Determine the temperature at which the pressure of each gas is zero.
   (ii) **A** is known to contain 0.024 mol of gas. Determine the number of moles of gas in **B**.

11 A balloon contains 0.050 m³ of hydrogen at an atmospheric pressure of $1.01 \times 10^5$ Pa and a temperature of 300 K. The balloon is taken to a depth of 50 m in sea water that has a temperature of 285 K and a density of 1100 kg m⁻³. Determine the volume of the balloon at this depth.

# Molecular Kinetic Theory

## Solids, Liquids and Gases

The kinetic theory is a mathematical model that explains macroscopic phenomena (phenomena concerned with bulk matter) in terms of the microscopic behaviour of small particles (atoms and molecules).

The idea that matter consists of atoms was proposed by the Greeks many centuries ago. In the 19th century, chemistry provided more hard evidence for such a model. For example: the fact that water ($H_2O$) always consists of the same mass of hydrogen combined with the same mass of oxygen however it is made.

In a **solid** the atoms are held together in a lattice as shown in Figure 4.59. They are able to vibrate but not to move relative to each other.

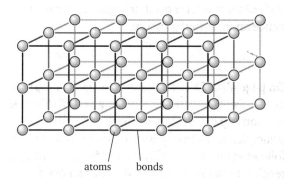

atoms    bonds

**Figure 4.59**

In a **liquid** the atoms (or molecules) are able to move around relative to each other. The molecules are close together so that the forces between them are large enough to keep them within the same volume and hinder their ability to escape from the surface of the liquid. The process of evaporation shows that some molecules can leave the surface.

In a **gas** the atoms and molecules are able to move around even more freely. They collide with each other and with the walls of the vessel that contains them. The atoms or molecules are further apart than in a liquid so the forces between them are small. The small forces result in the gas occupying the entire volume of the container it is in.

## The Molecular Kinetic Theory

In this model an ideal gas is considered to consist of a very large number of small molecules that behave like hard spheres. These move randomly within the container, colliding with one another and with the walls of the containing vessel.

The **pressure of the gas** is caused by the change in their momentum when they collide with the walls of the container. A collision of a molecule with the wall produces a force on the molecule and an equal and opposite force on the container.

A simple view of the situation is shown in Figure 4.60.

molecules bounce off the wall of the container

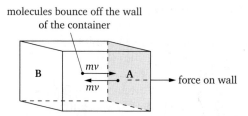

**Figure 4.60**

The following explains qualitatively how the pressure is produced.
- A molecule hits a wall **A** and reverses its velocity and hence its momentum.
- The molecule travels to the opposite wall **B** and travels back to wall **A** again.
- Each time it hits wall **A** it changes momentum and exerts a force on **A**.
- A very large number of molecules is doing this each second.
- The force is the total change in momentum each second, due to collisions of the molecules with wall **A**. ($F = \dfrac{\Delta(mv)}{t}$, see page 37)
- The pressure is equal to the force produced divided by the area of the surface with which the molecules collide ($p = F/A$).

## Assumptions of the kinetic theory

The assumptions of the molecular kinetic theory are:
- The volume of a molecule is small compared with the volume occupied by the gas.

- The forces of attraction between the molecules are negligible. The molecules only influence one another during a collision.
- The time between collisions is much greater than the duration of a collision.
- The collisions between molecules and collisions with the walls of the vessel are elastic.

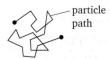

The theory also assumes that the gas consists of a very large number of molecules of the same mass moving in random directions at different speeds.

### Evidence for a molecular model

Evidence for the model is provided by:
- Brownian motion
- the success of the theory in explaining the behaviour of gases.

### Brownian motion

Brownian motion was first observed by Robert Brown. He noticed the rapid irregular motion of small pollen particles when they were suspended in water. A sketch of this motion is shown in the diagram in Figure 4.61.

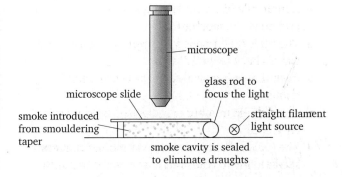

**Figure 4.61**

Using the arrangement shown in Figure 4.62, a similar irregular movement can be observed when smoke particles are suspended in air.

**Figure 4.62**

The kinetic theory explains this as follows.

The smoke particles are continually being bombarded on all sides by molecules of air. At any instant more air molecules may be hitting one side of the smoke particle than are hitting the other. The molecules striking each side may also have different speeds. This means that the change in momentum due to the collisions produces a greater force on one side than on the opposite side. The resultant force

produces movement in the direction of the force. A short time later the resultant force will be in a different direction, thus changing the direction of movement.

When a smoke particle is large, many more particles hit each side and the difference between the forces on opposite sides becomes relatively small so that no irregular movement is observed.

### Explaining why $p \propto \frac{1}{V}$ at constant temperature

When volume decreases:
- the molecules have less distance to travel before they collide with the sides of the container
- there are more collisions per second
- the change in momentum per second is increased
- the force and hence the pressure increases.

### Explaining why $p \propto T$ when volume is constant

When temperature increases, the speeds of the molecules increase. Therefore:
- there are more collisions per second with the walls of the containing vessel
- there is a greater change in momentum for each collision.

Both of these effects result in a greater force and hence an increased pressure.

### Derivation of $pV = \frac{1}{3} Nm\langle c^2 \rangle$

On page 46 the kinetic theory was used to explain qualitatively how pressure is produced. The above relationship can be derived using the same principles as in the qualitative explanation. The following derivation is very simplified, but the same result is obtained by a more rigorous analysis.

In the formula:
   $p$ is the pressure
   $V$ is the volume
   $N$ is the number of molecules in the gas
   $m$ is the mass of a molecule
   $\langle c^2 \rangle$ is the mean square speed of the molecules.

(**Note:** You will not be required to reproduce the following derivation but should understand the principles involved. You may be required to use the formula.)

For simplicity, suppose that:
- a gas is contained in a rectangular box of side lengths $a$, $b$ and $c$ as shown in Figure 4.63
- the gas contains $N$ molecules
- the molecules of the gas are all moving parallel to one of the three sides with a speed $v$.

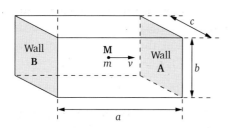

**Figure 4.63**

Consider one molecule **M** of mass $m$ moving at a speed $v$ towards the wall **A**, as shown in Figure 4.63.

The momentum of **M** towards the wall $= mv$.

When **M** hits wall **A** it rebounds with a speed that can be assumed equal to the speed before impact (since the container has a considerably larger mass than the molecule).

Its momentum after the collision with **A** $= -mv$.

The change in momentum $= mv - (-mv) = 2mv$.

This molecule will collide with **A** again when it has travelled to **B** and back to **A**. This distance is $2a$.

The time taken to travel $2a = 2a/v$.

The number of impacts per second of **M** with

wall $\mathbf{A} = \dfrac{1}{2\dfrac{a}{v}} = \dfrac{v}{2a}$

The total momentum change per second of **M**

during impacts with wall $\mathbf{A} = 2mv\,\dfrac{v}{2a}$

Change in momentum per second is equal to the force so this is the force exerted by **M** on wall **A**.

There are $N$ molecules altogether so it is assumed that on average $\dfrac{N}{3}$ of these will be moving parallel to each side.

The total force $F$ = total change in momentum per second on the wall)   $\mathbf{A} = 2mv\,\dfrac{v}{2a} \times \dfrac{N}{3} = \dfrac{Nmv^2}{3a}$

The pressure on wall  $\mathbf{A} = \dfrac{F}{\text{Area of } \mathbf{A}} = \dfrac{F}{bc} = \dfrac{Nmv^2}{3abc}$

where $abc$ = volume of the gas.

Hence $p = \dfrac{1}{3}\dfrac{Nmv^2}{V}$ or $pV = \dfrac{1}{3}Nmv^2$.

The assumption that all molecules move at the same speed is not correct. Molecules in a gas have a range of speeds as shown by the graph in Figure 4.64.

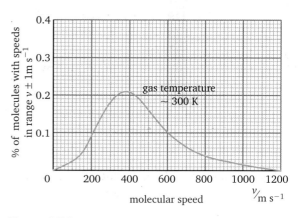

**Figure 4.64**

As all collisions are elastic, kinetic energy is exchanged when two molecules collide, but since the collisions are perfectly elastic the total kinetic energy remains the same.

Using the range of speeds, a more rigorous analysis shows that the **mean square speed** should be used instead of $v^2$. The mean square speed is the mean of the squares of the speeds of the molecules and is written mathematically as $\langle c^2 \rangle$.

The relevant speed of the molecules is then called the **root mean square speed** $\sqrt{\langle c^2 \rangle}$.

### Effect of temperature on the mean square speed

When temperature rises, the total kinetic energy of the molecules increases.

The graphs in Figure 4.65 show how the speeds of the molecules in a gas change when temperature increases.

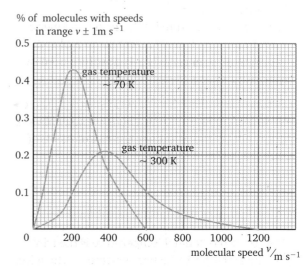

**Figure 4.65**

## Relationship between kinetic energy and temperature

We now have two formulae for the product $pV$:

- one from the ideal gas equation (see page 44)
$$pV = nRT$$
- and one from the kinetic theory
$$pV = \tfrac{1}{3}Nm\langle c^2 \rangle$$

Equating the two right hand sides gives:
$$\tfrac{1}{3}Nm\langle c^2 \rangle = nRT$$

$$\tfrac{2}{3}N\tfrac{1}{2}m\langle c^2 \rangle = nRT$$

If there is 1 mol of gas then this becomes:
$$\tfrac{2}{3}N_A\left(\tfrac{1}{2}m\langle c^2 \rangle\right) = RT$$

$\left(\tfrac{1}{2}m\langle c^2 \rangle\right)$ is the average kinetic energy $E_k$ of a molecule, so from the previous equation:
$$\langle E_k \rangle = \frac{3}{2}\frac{R}{N_A}T = \frac{3}{2}kT$$

$$\frac{R}{N_A} = k, \text{ the Boltzmann constant}$$

$R = 8.3 \text{ J mol}^{-1}\text{ K}^{-1}$ and $N_A = 6.0 \times 10^{23}\text{ mol}^{-1}$

$$k = \frac{8.3}{6.0 \times 10^{23}} = 1.38 \times 10^{-23}\text{ J K}^{-1}$$

## Summary

- The absolute temperature $T$ is proportional to the mean kinetic energy of all the molecules in the gas.
- The average kinetic energy of a gas molecule increases by $1.38 \times 10^{-23}$ J for every kelvin rise in temperature.
- When the temperature is 0 K, the mean kinetic energy of the molecules is zero so no molecule has any kinetic energy. This is **absolute zero of temperature (–273°C)**.

At absolute zero there is no kinetic energy so there is no molecular momentum. There are no collisions with the wall of the container and the pressure of the gas becomes zero.

## Thermodynamic Scale of Temperature

On this scale the defined temperature is the triple point of water. This is the temperature at which ice, water and water vapour exist together in equilibrium. The temperature at the triple point of water is defined as 273.16 K.

Other temperatures are then given by

$$T/K = \frac{(pV)_T}{(pV)_{tr}} \times 273.16$$

If the volume of a container is kept constant this becomes:

$$T/K = \frac{(p)_T}{(p)_{tr}} \times 273.16$$

This is the definition of temperature using a constant volume gas thermometer.

## Zeroth Law of Thermodynamics

This is concerned with the meaning of temperature. One body **A** is said to be at a higher temperature than another **B** if, when they are placed in contact, energy flows from **A** to **B**.

Using the term 'hotter' can lead to problems as hotness, like loudness, is subjective. If you touch a sheet of steel and a piece of rubber which are at the same temperature the rubber appears hotter. This is because the rate of transfer of energy to the steel from the body is quicker than to the rubber at the same temperature. If the rubber and steel were placed in contact, there would be no net exchange of energy.

If no net energy exchange takes place when two bodies are placed in contact, the bodies are at the same temperature. The bodies are then said to be in **thermal equilibrium**. Stated formally, the Zeroth law puts this another way.

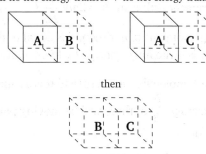

Figure 4.66

If no net energy transfer + no net energy transfer

then

no net energy transfer

**Figure 4.66**

If, in Figure 4.66:
- **A** is in thermal equilibrium with **B**
- **A** is also in thermal equilibrium with **C** then
- **B** is in thermal equilibrium with **C**.

Taking a more practical view, suppose you take the temperature of two different solid objects. If the thermometer reads the same value for both objects there would be no change in the temperature of either object when they are placed in contact.

If one had a higher temperature, energy would flow from that body to the other. Its temperature would fall and the temperature of the other body would rise.

## Zeroth Law on a Microscopic Scale

The atoms of an object at a higher temperature have a higher mean kinetic energy.

At the interface between a high temperature object and a low temperature object, collisions are taking place between the atoms.

During the collisions energy is transferred from the atoms with higher kinetic energy to those that have lower energy.

There will be more molecules with higher kinetic energy in the body at the higher temperature.

The mean kinetic energy of the atoms in the object that is at the higher temperature therefore falls. This continues until the atoms in each object have the same mean kinetic energy, in other words, until they have the same temperature.

**Test your understanding**

1 Calculate the change in momentum when a molecule of mass $3.4 \times 10^{-27}$ kg travelling at $450$ m s$^{-1}$ rebounds elastically from the wall of a containing vessel.

2 The pressure in the vessel in question 1 is $1.2 \times 10^5$ Pa. Calculate the number of molecules that strikes each cm$^2$ of the wall each second.

3 Calculate the number of molecules of mass $3.4 \times 10^{-27}$ kg and root mean square (rms) speed $510$ m s$^{-1}$ in a sample that is at a pressure of $1.01 \times 10^5$ Pa and occupies a volume of $0.10$ m$^3$.

4 (a) Calculate the kinetic energy of a molecule at a temperature of:
 (i) 300 K
 (ii) 1000 K.
 (b) A molecule of oxygen has a mass of $5.3 \times 10^{-26}$ kg. Calculate its rms speed at temperatures of 300 K and 1000 K.
 $k = 1.38 \times 10^{-23}$ J K$^{-1}$.

5 Calculate the rms speed of helium atoms (mass $6.8 \times 10^{-27}$ kg) at a temperature of 10 K.

6 A molecule will escape from the Earth when its speed is $11\,000$ m s$^{-1}$.
 (a) Calculate the temperature at which the rms speed of carbon dioxide molecules of mass $7.5 \times 10^{-26}$ kg will reach this speed.
 (b) Explain why carbon dioxide molecules will escape at temperatures well below that which you have calculated in (a).

7 (a) Which gas has the higher temperature? **A**, a gas with molecules of mass $6.8 \times 10^{-27}$ kg travelling at a rms speed of $300$ m s$^{-1}$, or **B**, a gas with atoms of mass $3.4 \times 10^{-26}$ kg with a rms speed of $200$ m s$^{-1}$.
 (b) Calculate the temperature difference between **A** and **B**.

# Heating and Working

## Internal Energy

Internal energy is the random distribution of potential and kinetic energy amongst the atoms or molecules of an object.

The molecules in a solid can vibrate in their lattice so that each molecule possesses energy that is changing between kinetic and potential energy. The energy can be passed from one atom or molecule to another, but the total energy will remain constant unless energy is supplied to or taken away from the system.

In liquids and gases the molecules have translational (or linear) kinetic energy and so move around randomly. The kinetic energy of a molecule will vary as it moves due to collisions with other molecules. Unless energy is supplied or extracted, the total kinetic energy of the molecules, and hence the mean kinetic energy of a molecule, will be constant.

Supplying energy to a substance increases the internal energy.

When the substance is a solid, liquid or gas, the internal energy change is in the form of kinetic energy and the temperature rises.

When a substance changes from solid to liquid or from liquid to gas, the internal energy change is in the form of potential energy. The energy input frees or separates the molecules. During these processes the temperature remains constant since there is no change in kinetic energy.

The two processes that produce a change in internal energy of an object are:
- **heating**
- **working**.

## Heating

Heating is the process that occurs when the internal energy of a system is increased due to energy transferred to it from a system that is at a higher temperature.

In Figure 4.67, object **A** is in contact with **B**, which is at a higher temperature.

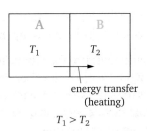

energy transfer
(heating)

$T_1 > T_2$

**Figure 4.67**

The average kinetic energy of the molecules in **B** is greater than that in **A**. The result is that:
- molecules in **B** transfer energy to the molecules in **A**
- the total internal energy of **A** increases while that of **B** decreases
- the mean kinetic energy of the molecules in **A** increases and that in **B** decreases
- the temperature of **A** rises and that of **B** falls.

### Examples of heating

When using a Bunsen burner, the thermal energy (heat) flows from the hot gases to the colder object.

When using a saucepan on an electric cooking ring, the energy flows from the element that is at a higher temperature than the saucepan.

When bread is placed in a freezer compartment, energy flows from the bread to the refrigerator, which is at a lower temperature.

The Earth is heated by the Sun due to energy transfer by photons of electromagnetic radiation. All bodies that are above absolute zero radiate energy. If energy were not received from the Sun the temperature of the Earth would fall because it would radiate energy.

The Earth is in a state of **thermal equilibrium**. The heating of the Earth due to the Sun is equal to the cooling of the Earth due to its emitted radiation.

## Working

Working is the process of changing the internal energy of an object by doing mechanical work.

For example:
- continually hitting a metal block with a hammer
- rubbing one metal against another that is at the same temperature.

In these two processes, work is done and there is a rise in temperature of both objects due to an increase in their internal energy.

Note that there is no heating process taking place although the temperature is rising. The increase in internal energy comes from the person doing the hammering or from the energy source that is causing the metals to move relative to one another.

An important example of changing internal energy by doing work is the compression or expansion of a gas. This is discussed in detail on page 53.

## First Law of Thermodynamics

This states that the change in internal energy of a system (a solid, liquid or gas) is the sum of the heating and working that has taken place.

A mathematical statement of the first law of thermodynamics is:

$$\Delta U = Q + W$$

In this equation:
- $\Delta U$ is the **change in internal energy** of the system
- $Q$ is the **thermal energy (or heat) input** to the system
- $W$ is the **work done on** the system.

This is shown diagrammatically in Figure 4.68.

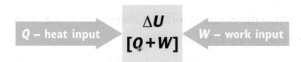

**Figure 4.68**

If work is done **by** the system or if there is a flow of heat **from** the system the signs of the quantities $Q$ and $W$ are changed.

**EXAMPLE ONE:** A system has 250 J of work done on it and has 150 J of heat supplied from a body that is at a higher temperature.

The total change in internal energy is given by:

$$\Delta U = + 150 + 250 = 400 \text{ J}$$

**EXAMPLE TWO:** 500 J of work is done on a sheet of metal by hammering it. During the time it is being hammered, it transfers 300 J to the surroundings.

The overall change in internal energy of the metal sheet and the hammer is given by:

$$\Delta U = -300 + 500 = 200 \text{ J}$$

## An Isothermal Change

An isothermal change takes place with no change in temperature. Time has to be allowed during the change for the temperature to remain steady.

Since the temperature is constant, the internal energy of the system is unchanged.

$$\Delta U = 0$$

If work is done **on** the system there must be an equal amount of heat lost **from** the system.

$$\Delta U = 0 = W + Q$$
$$W = -Q$$

## An Adiabatic Change

An adiabatic change takes place quickly so that there is no time for the system to lose energy to the surroundings.

$$Q = 0$$

and

$$\Delta U = W$$

An adiabatic change in which work is done **on** the system (i.e. the system is compressed) will increase the internal energy and raise the temperature.

An adiabatic change in which work is done **by** the system (i.e. the system expands) will decrease the internal energy and lower the temperature.

## Change at Constant Volume

In a constant volume change there is no work done either by or on the system.

$$W = 0$$

This means that:

$$\Delta U = Q$$

The increase in internal energy in this case is all due to energy transfer to the system from one that has a higher temperature.

The heating of a solid or liquid is, more or less, a constant volume change. When a solid is heated there is only a very small change in volume due to expansion. There is a relatively small amount of work done in pushing back the surroundings.

Although liquids expand more, the work done is still small compared with the energy needed to change the kinetic energy of the molecules.

When a gas is heated in a cylinder fitted with a piston the heating will only take place at constant volume if the piston is held in place. If the piston moves due to the increased pressure of the gas, the gas does work on the surroundings.

**Test your understanding**

1  Complete the following table.

**Table 4.3**

|   | $\Delta U$/kJ | $Q$/kJ | $W$/kJ |
|---|---|---|---|
| A | 200 | 100 | |
| B | | 50 | −50 |
| C | | −300 | 250 |
| D | 120 | | −400 |

2  In which of the changes in question **1**:
   **(a)** does the temperature of the system fall?
   **(b)** does the system do work on the surroundings?
   **(c)** is the system heated?

# Work Done on or by a Gas

To compress a gas, a force is needed. The force will move its point of application as the gas compresses, so work will be done on the gas.

Consider a gas held by a piston in a sealed cylinder (such as a bicycle pump with a finger held over its outlet) as shown in Figure 4.69.

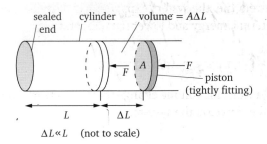

$\Delta L \ll L$    (not to scale)

**Figure 4.69**

When the piston is pushed in, the force $F$ needed to move the piston is approximately equal to the force on the piston due to the pressure $p$ of the gas.

$$F = pA$$

where $A$ is the area of cross-section of the piston, shown in Figure 4.69.

For a change at constant pressure (i.e. when the distance $\Delta L$ is small) the work done will be given by:

$$W = F\Delta L$$

$$\therefore\ W = pA\Delta L$$

where $A\Delta L = \Delta V$, the change in volume of the gas.

The work done on the gas when compressing it at constant pressure is therefore given by:

$$W = p\Delta V$$

For an **adiabatic change** (when no energy is lost to the surroundings) this will be equal to the change in internal energy of the gas.

In an **isothermal change** (when temperature is constant) the energy input due to work done on the gas has to be lost to heat the surroundings.

## Use of p-V graphs

Figure 4.70 shows a p-V graph for a fixed mass of gas at constant pressure.

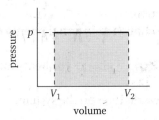

**Figure 4.70**

When the volume **increases** from $V_1$ to $V_2$ the work done **by** the gas is given by:

$$W = p\,(V_2 - V_1)$$

This is the area under the graph between the line and the volume axis.

The area under the graph line for any variation of $p$ with $V$ gives the work done.

A volume **decrease** would result in work done **on** the gas.

Figure 4.71 shows the *p-V* diagram for an isothermal change.

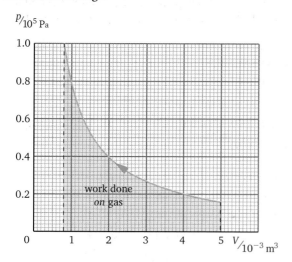

**Figure 4.71**

- The arrow on the line signifies that the gas is being compressed.
- The shaded area is the work done on the gas.

The work done can be estimated by:
- counting the number of squares under the graph line
- calculating the work represented by one square
- multiplying these together.

Note that the whole area must be determined down to $p = 0$ so beware when the pressure axis uses a false origin!

## Indicator diagrams

In devices such as engines and refrigerators, gases go through cycles of pressure–volume changes. The cycle of changes is called an indicator diagram. An example of a simple indicator diagram for an engine is shown in Figure 4.72.

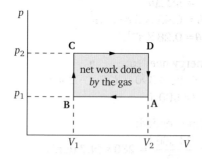

**Figure 4.72**

- From **A** to **B** the gas is compressed at constant pressure. The work done **on** the gas $= p_1(V_2 - V_1)$.
- From **B** to **C** the pressure increases at constant volume. This would be achieved by heating the gas. No work is done either on or by the gas since there is no volume change.

- From **C** to **D** the gas expands at constant pressure. The work done **by** the gas is $p_2(V_2 - V_1)$.
- From **D** to **A** the pressure falls at constant volume. This would be achieved by cooling the gas. No work is done either on or by the gas since there is no volume change.

In one cycle of changes net work is done **by** the gas **on** the surroundings since $p_2$ is greater than $p_1$.

The net work is $(p_2 - p_1)(V_2 - V_1)$.

This is the area enclosed by the *p-V* graph.

Figure 4.73 shows this diagrammatically.

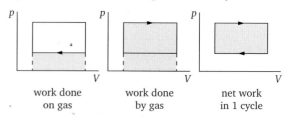

| work done on gas | work done by gas | net work in 1 cycle |

**Figure 4.73**

The shaded areas in graphs **1**, **2** and **3** show respectively:
- the work done **on** the gas during one cycle
- the work done **by** the gas during one cycle
- the net work done **by** the gas **on** the surroundings during the cycle.

This net work could be used to drive a mechanical system such as an engine.

Real engines have much more complex cycles of changes. A simplified version of a more typical cycle is shown in Figure 4.74.

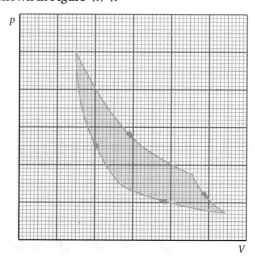

**Figure 4.74**

The net work done per cycle is found by determining the work represented by the enclosed area of the graph.

In a refrigerator, the direction of the cycle is reversed so that there is net energy input to the system during a cycle. This energy input comes from the cooling of the items that are in the refrigerator.

## Specific Heat Capacity

The specific heat capacity $c$ is the energy needed to raise the temperature of 1 kg of a substance by 1 K.

Therefore, the heat input $Q$ needed to raise the temperature of an object of mass $m$ by $\Delta\theta$ is given by:

$$Q = mc\,\Delta\theta$$

Since $c = \dfrac{Q}{m\,\Delta\theta}$ the unit of specific heat capacity is $\text{J kg}^{-1}\,\text{K}^{-1}$.

The specific heat capacity depends on the substance. The values for some substances are given below.

Water 4200 J kg$^{-1}$ K$^{-1}$

Copper 390 J kg$^{-1}$ K$^{-1}$

Aluminium 910 J kg$^{-1}$ K$^{-1}$

Concrete 3400 J kg$^{-1}$ K$^{-1}$

Sodium 1200 J kg$^{-1}$ K$^{-1}$

The higher the heat capacity the more energy is needed to raise the temperature by 1 K.

The definition usually assumes that the energy input is due to heating. However, if there is no change in volume so that no work is done **by** the system the energy input needed to raise the temperature by 1 K is the same whether the energy input is due to heating or working.

EXAMPLE: A 50 W electric heater is embedded in an insulated block of copper of mass 0.450 kg.

(a) Calculate the temperature rise in one second.

(b) How long would it take to change the temperature from 290 K to 370 K?

(a) Energy input per second = 50 J
$Q = mc\,\Delta\theta$
$50 = 0.45 \times 390 \times \Delta\theta$
$\Delta\theta = 0.28\text{ K s}^{-1}$.

(b) Energy needed:
$= 0.45 \times 390 \times (370 - 290)$
$= 14\,000$ J.

Time needed:
$= \dfrac{14\,000}{50} = 280$ s (4.7 min).

## Continuous Flow Heating

Continuous flow heating is very commonly used in practical applications.

When a fluid (a liquid or gas) is passed over a surface that has a higher temperature, as shown in Figure 4.77 the fluid is heated.

1 Calculate the work done when a gas expands from a volume of 0.0015 m³ to 0.0035 m³ at a constant pressure of $8 \times 10^4$ Pa. State whether work is done on or by the gas.

2 Determine the work done in each of the following instances in Figure 4.75. In each case state whether work is done on or by the gas.

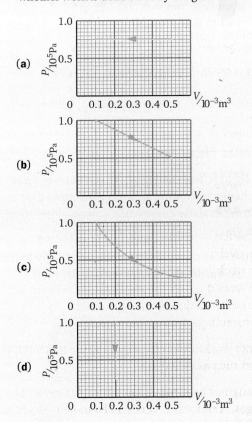

**Figure 4.75**

3 Figure 4.76 shows the indicator diagram for a fixed mass of gas.

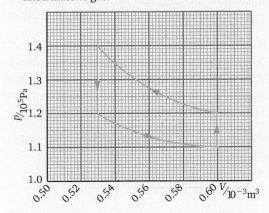

**Figure 4.76**

(a) Determine the net work done during one cycle.

(b) State whether the work results in an energy input or output.

(c) What type of machine would operate using such a cycle of changes?

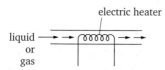

**Figure 4.77**

Suppose that:
- the rate at which a fluid flows past a heater is
$$\frac{\Delta m}{\Delta t}$$
- the specific heat capacity of the fluid is $c$
- the temperature of the fluid before being heated is $\theta_1$
- the temperature afterwards is $\theta_2$.

The energy that is transferred to the fluid per second $\frac{\Delta Q}{\Delta t} = \frac{\Delta m}{\Delta t} c \left( \theta_2 - \theta_1 \right)$.

This energy has been extracted from the heater surface. The heater will cool unless energy is supplied to it at the same rate as energy is transferred to the fluid.

In some applications the system is designed to heat fluid that is used for a particular purpose.

For example, in **central heating systems** the fluid is water and the energy is used to heat radiators. In **nuclear power stations** the fluid may be water or a gas. The fluid flows through the hot reactor core and is heated. The hot fluid can then be used to generate steam to drive a turbine that, in turn, powers a generator.

In other applications the purpose of the cooling fluid is to maintain the system that is heating the fluid at a constant temperature although energy is continually supplied to it.

For example, in a **car engine** a considerable amount of energy is generated that is not used to drive the car. This raises the temperature of the car engine. Unless removed this heat will quickly damage the engine. Water (or in some cases air) passing through the engine maintains it at a suitable temperature. This energy is transferred to the surroundings by the continuous heating of air that is drawn through the radiator by a fan. Similarly, a high temperature will damage a **computer**. A fan is used to drive a flow of air over the processor to maintain a suitable temperature.

EXAMPLE: The energy that has to be removed from a computer to maintain a suitable working temperature is 50 W. The temperature of the air entering the computer is 30°C. The temperature of the air leaving the computer vents is 45°C. Calculate the volume of air that has to pass through the computer each second to maintain this temperature.

Density of air = 1.3 kg m$^{-3}$.
Specific heat capacity of air = 990 J kg K$^{-1}$.

Energy transferred to air per second = 200 J.
$$\frac{\Delta Q}{\Delta t} = \frac{\Delta m}{\Delta t} c \left( \theta_2 - \theta_1 \right)$$
$$50 = \frac{\Delta m}{\Delta t} \times 990 \times \left( 45 - 30 \right)$$

Mass flow rate = 0.0037 kg s$^{-1}$

$$\text{Volume flow rate} = \frac{\text{mass flow rate}}{\text{density}}$$
$$= \frac{0.0037}{1.3}$$
$$= 2.6 \times 10^{-3} \text{ m}^3 \text{ s}^{-1}$$

## Specific Heat Capacity of Gases

In many practical situations a fixed mass of a gas may change in volume as energy is supplied or extracted. In these cases work is done either by or on the gas.

In Figure 4.78 energy is supplied electrically to the gas in the insulated cylinder.

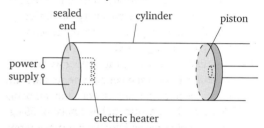

**Figure 4.78**

When the piston remains in the same place so that the volume is unchanged, all the energy is used to change internal energy and raise the temperature of the gas. The pressure of the gas increases.

If the piston is now allowed to move so that the gas expands to maintain a constant pressure, work is done ($= p \, \Delta V$) **by** the gas. There is a fall in internal energy due to this work and the temperature of the gas falls.

This means that the energy input needed to raise the temperature of the gas by 1 K at constant volume is less than that required to produce a 1 K rise at constant pressure.

Therefore, the specific heat capacity of a gas at constant pressure is greater than that at constant volume ($c_p > c_v$).

The constant pressure and constant volume conditions are two particular cases. The specific heat capacity of a gas can take any value, depending on how much the gas expands during the heating.

## Test your understanding

1  How much energy is required to raise the temperature of a 1.5 kg iron bucket of specific heat capacity 450 J kg$^{-1}$ K$^{-1}$ from 20°C to 45°C?

2  How much energy would be needed to raise the temperature of the same bucket as in **1** through the same temperature rise when it contains 3.5 kg of water?

3  Calculate the specific heat capacity of a metal that requires 270 kJ of heat to raise 4.8 kg of the metal from 290 K to 310 K.

4  An electrical heater supplies energy to 1.5 kg of water in a kettle at a rate of 2.0 kW. Calculate the time taken to raise the temperature from 20°C to 100°C, assuming that negligible energy is required to raise the temperature of the kettle itself.

5  In a central heating system 35 kg of water flows through the boiler each minute. The water temperature increases by 2.5 K each time it passes through the boiler. Calculate the useful thermal output of the boiler.

6  A 0.45 kg lump of lead is dropped from the top of a building 100 m high. The specific heat capacity of lead is 126 J kg$^{-1}$ K$^{-1}$.

   (a)  Assuming that all the gravitational potential energy becomes internal energy in the lead, calculate the temperature rise of the lead when it hits the ground.

   (b)  Give two reasons why the temperature rise would be lower than your answer to part (a).

7  The mean specific heat capacity of a human being is 3500 J kg$^{-1}$ K$^{-1}$. A student has a mass of 65 kg and during a physical activity the internal energy of the student rises at a rate of 250 W.

   Supposing that no energy were lost from the student to the surroundings due to the body's cooling mechanisms, calculate the time taken for the body temperature to rise from 37°C to a critical body temperature of 43°C.

## Changing Phase

There are four **phases** of matter. These are solid, liquid, gas and plasma. The plasma phase is only of concern at extremely high temperatures such as those found in stars and nuclear fusion reactors. The plasma phase is not studied in this unit.

Note that the phases of matter are sometimes referred to as states of matter. This is ambiguous, however, since the state of a system such as a gas (see page 42) can be changed by changing temperature, pressure or volume.

## The Specific Latent Heat of Fusion

When a substance is melting (or fusing) the energy supplied does not raise the temperature. All the energy supplied is used to break intermolecular bonds and change the phase. The energy input becomes internal potential energy.

The specific latent heat is the energy needed to change 1 kg of the substance from solid to liquid when the solid is at its melting point.

The energy required to change $m$ kg of solid to liquid is given by:

$$Q = ml_f$$

where $l_f$ is the specific latent heat of fusion.

## The Specific Latent Heat of Vaporisation

When the boiling point is reached, once again the energy supplied does not change temperature but is used to free molecules from the liquid to become gas (or vapour).

The specific latent heat of vaporisation is the energy needed to change 1 kg of the substance from liquid to gas when the liquid is at its boiling point.

The energy required to change $m$ kg of solid to liquid is given by:

$$Q = ml_v$$

where $l_v$ is the specific latent heat of vaporisation.

The specific latent heat of vaporisation of a substance is always much greater than the specific latent heat of fusion.

**Table 4.4**

**Some values for specific latent heats**

| Substance | $l_f$/kJ kg$^{-1}$ | $l_v$/kJ kg$^{-1}$ |
|---|---|---|
| water | 330 | 2300 |
| copper | 200 | 5100 |
| lead | 23 | 870 |
| silver | 110 | 2400 |
| hydrogen | 59 | 450 |

EXAMPLE: A 3.0 kW kettle starts to boil. Assuming that all the energy becomes latent heat in the steam that is produced, calculate the mass of water that evaporates in 5.0 minutes.

Energy supplied in 5.0 minutes = 3000 × 5 × 60
= 900 kJ

$$Q = ml_v$$
$$9.0 \times 10^5 = m \times 2.3 \times 10^6$$
$$m = 0.39 \text{ kg}$$

## Molar heat capacity and molar latent heat

It is sometimes more convenient to relate heat capacity and latent heat to the number of moles of the substance in a sample rather than to its mass in kg.

The definitions are then as follows.
The molar heat capacity $C$ is the energy needed to raise the temperature of 1 mol of a substance by 1 K. The unit is J mol$^{-1}$ K$^{-1}$.

The molar latent heat of vaporisation $L_v$ is the energy needed change 1 mol of the substance from liquid to gas when the liquid is at its boiling point. The unit is J mol$^{-1}$.

The molar latent heat of fusion $L_f$ is the energy needed change 1 mol of the substance from solid to liquid when the solid is at its melting point.

The formulae for calculating the energy required to raise the temperature of $n$ mol by a temperature $\Delta\theta$ becomes

$$Q = n\,C\,\Delta\theta$$

For latent heat the formula becomes

$$Q = n\,L$$

where $L$ is the appropriate molar latent heat.

You may need to convert from specific to molar values. The molar heat capacity and molar latent heat are found by multiplying the corresponding specific heat capacity or specific latent heat by the molar mass in kg.

For example:
The specific heat capacity of water is 4200 J kg$^{-1}$ K$^{-1}$. The molar mass of water is 0.018 kg. The molar heat capacity of water is 75.6 J mol$^{-1}$ K$^{-1}$.

The energy required per molecule of water can now be calculated using the Avogadro constant.

### Test your understanding

1  Using values from Table 4.4 calculate:
   (a) calculate the energy needed to melt 1.3 kg of silver when it is at its melting point
   (b) the energy needed to vaporise 1.5 kg of water at 100°C
   (c) the power needed to melt ice at a rate of 0.0005 kg s$^{-1}$.
2  A saucepan is on a cooker ring that has a power of 800 W.
   (a) Calculate the minimum time taken to evaporate 0.50 kg of water in a saucepan when the water initially has a temperature of 20°C.
   (b) The time taken would actually be longer than your answer to (a). Explain why.

3  A forensic scientist notices that a lead bullet of mass $5.5 \times 10^{-3}$ kg became embedded in a table and melted completely on impact. Calculate the minimum velocity of the bullet on impact assuming that it was at room temperature (300 K) when it struck the table.
   (Specific heat capacity of lead = 130 J kg$^{-1}$ K$^{-1}$, melting point of lead = 600 K.)
4  A student's body temperature remains constant at 39°C when thermal energy is generated in the body at a rate of 200 W during a run. Assuming that the heat is lost through evaporation of water from the surface of the body, calculate the mass of water that evaporates in 10 minutes.

## Working on Wires and Rods

Provided that a spring obeys Hooke's law ($F \propto \Delta l$), the energy transferred when a force $F$ applied to a spring causes it to extend by a length $\Delta l$ is given by

$$E = \tfrac{1}{2}k\Delta l^2$$

This is also the work done on the spring (see page 24).

When a force is applied to stretch a wire, most wires initially obey Hooke's law, and this same equation gives the work done.

A solid rod initially obeys Hooke's law when extended or compressed, and the formula can be used in each case.

The stiffness of a given wire can be calculated as for a spring using the formula $F = k\,\Delta l$. The stiffness $k$ of a wire depends on:
● the material from which it is made
● the area of cross section $A$ of the wire
● the initial length $l$ of the wire.

**Young's modulus** is a property of the wire that takes account of all these variables.

## Stress, Strain and Young's Modulus

Figure 4.79 shows a wire being stretched.

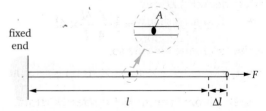

**Figure 4.79**

$F$  is the force applied
$A$  is the area of cross-section of the wire
$\Delta l$  is the extension produced by the force
$l$  is the length of the wire before the force was applied (the initial length).

We say that the wire has been subjected to a **stress** and that this produces a **strain**.

**Stress** $\sigma$ is defined as the force per unit area.

$$\sigma = \frac{F}{A}$$

The unit of stress is $N\,m^{-2}$ or pascal Pa.

**Strain** $\varepsilon$ is defined as the extension per unit length.

$$\varepsilon = \frac{\Delta l}{l}$$

Strain is a ratio of two lengths and therefore it has no unit.

For a wire that obeys Hooke's law, which most wires do for small extensions, the stress is proportional to the strain ($\sigma \propto \varepsilon$) so that the ratio of stress to strain is a constant for that wire. The maximum stress for which stress is proportional to the strain is called the **limit of proportionality**.

**Young's modulus** $E$ for the material of a wire or rod is given by:

$$E = \frac{\text{stress}}{\text{strain}} = \frac{\sigma}{\varepsilon}$$

The unit of Young's modulus is the same as that of stress, $N\,m^{-2}$ or Pa.

Using the definitions of stress and strain:

$$E = \frac{F/A}{\Delta l/l} = \frac{Fl}{A\,\Delta l}$$

This is a useful formula. When any four of the variables are known, the other can be determined.

Remember that this formula must only be used for the initial straight part of a stress–strain graph. The graph for most materials will curve for higher forces.

## Elastic strain energy

Within the Hooke's law region where $F \propto \Delta l$, the total elastic stored energy $= \frac{1}{2} F\Delta l$

Writing this another way

$$\text{energy stored} = \frac{1}{2}\,\frac{F}{A}\,\frac{\Delta l}{l} \times Al$$

$Al$ is the volume of the wire so:

$$\text{energy stored} = \frac{1}{2} \times \text{stress} \times \text{strain} \times \text{volume}$$

or

$$\text{energy stored per } m^3 = \frac{1}{2} \times \text{stress} \times \text{strain}.$$

In the Hooke's law region this is the area under the stress–strain graph.

The strain energy is potential energy stored in stretched inter-atomic bonds. These will return to their original length when the stretching force is removed.

**Note:** For this course you should concentrate on the macroscopic (large-scale) behaviour of materials. A detailed understanding of the microscopic (small-scale) behaviour of materials when they are stressed is not needed. Brief explanations of the behaviour of materials in molecular terms are included for interest only.

EXAMPLE: A force of 100 N is applied to the end of a steel wire 2.5 m long and of diameter 1.5 mm. The Young modulus of steel is $2.1 \times 10^{11}$ Pa .

Calculate:

(a) the stress in the wire

(b) the extension of the wire

(c) the elastic strain energy stored in the wire.

(a)  Stress $= F/A$

$$A = \frac{\pi \times (1.5 \times 10^{-3})^2}{4} = 1.77 \times 10^{-6}\ m^2$$

$$\text{Stress} = \frac{100}{1.77 \times 10^{-6}} = 5.7 \times 10^7\ Pa$$

(b)
$$E = \frac{\text{Stress}}{\text{Strain}}$$

$$\text{Strain} = \frac{\text{Stress}}{E} = \frac{5.7 \times 10^7}{2.1 \times 10^{11}} = 2.7 \times 10^{-4}$$

$$\frac{\text{Extension}}{\text{original length}} = \text{Strain}$$

$$\text{Extension} = \text{Strain} \times \text{original length}$$

$$= 2.7 \times 10^{-4} \times 2.5 = 6.7 \times 10^{-4}\,m$$

(c)  Strain energy $=$

$\frac{1}{2}$ force $\times$ extension $=$

$\frac{1}{2} \times 100 \times 6.7 \times 10^{-4} = 0.034$ J

## Yielding and breaking stresses and strains

The **breaking stress (or strain)** is the stress (or strain) at which a sample of the material breaks.

The breaking stress is a property of the material and is often referred to as the **ultimate tensile stress**. The **breaking force** is a characteristic of a particular sample of the material and is the breaking stress multiplied by the area of cross-section of the sample.

For many materials the breaking stress is the important factor when deciding whether they are useful in designing a structure. However, different materials exhibit different behaviour after reaching the end of the linear region so this is not always the case.

**Brittle** materials such as cast iron, glass and concrete are linear up to the point at which they break (see Figure 4.80).

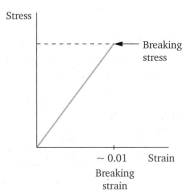

**Figure 4.80**

In brittle materials, small cracks that appear weaken the material due to **stress concentration** at the tip of the crack (see Figure 4.81).

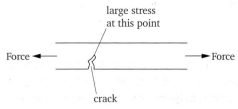

**Figure 4.81**

Continual flexing of the material causes the cracks to become larger. **Crack propagation** occurs and leads to a sudden breaking of the material. This has been the cause of many disasters in bridges and in transport, such as iron girders collapsing and rail-tracks breaking so that trains leave the tracks.

Materials such as copper, aluminium and steel are **ductile**, which allows them to be drawn into wires. When they are subjected to a particular stress they **yield** (see Figure 4.82) before they break.

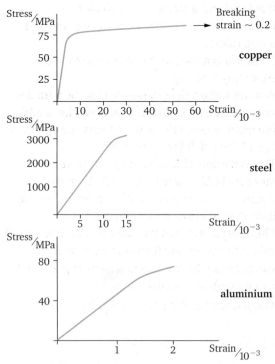

**Figure 4.82**

When the **yield stress** (or **yield strain**) is reached the material will:
- undergo large increases in strain for very small increases in stress
- not return to its original shape when the force is removed.

As Figure 4.82 shows, the strain introduced between yielding and breaking varies from material to material.

For this class of materials it is the yield stress rather than the breaking stress that is the important factor in the design of structures.

When yielding:
- the material is deformed **plastically**
- will no longer return to its original length when the force is removed
- the arrangement of atoms within the material is altered and is not reversible
- the work done in rearranging the atoms is not retrievable.

There may also be an increase in temperature of the wire due to an increase in its internal energy. This energy will be transferred to the surroundings by heating. This energy is therefore also irretrievable when the stretching force is removed.

### Stretching rubber

A typical stress–strain graph for stretching rubber in the form of a rubber band is shown in Figure 4.83.

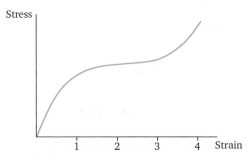

**Figure 4.83**

**Rubber** is an elastomer. This class of materials can be stretched to very large strains ($\approx 5$ or 500%) without breaking. It is hard to stretch at first, then it becomes easier, and then harder again.

The following explains why this occurs at a molecular level. Rubber consists of a mass of tangled, long-chain molecules. It is hard to stretch at first as tangled molecular chains are aligned.

The stretching becomes easier once the tangled molecules have been aligned. There is a large increase in strain for a small additional force. This is because the molecular chains are not straight but are at angles, rather like a concertina.

A relatively small force is needed to change the angles between the bonds.

Once the bonds are straight, the work done is now used to stretch the bonds so that the stretching becomes more difficult again.

### Work done when stretching materials

Many materials have an elastic region. This means that all the energy transferred to a material when stretching is retrieved when the tensile force is removed.

For metal wires the elastic behaviour occurs for a little way beyond the part where tensile stress in proportional to strain.

Beyond the **elastic limit**, the metal does not return to its original length but has a **permanent strain** when the force is removed. This is shown in Figure 4.84(a).

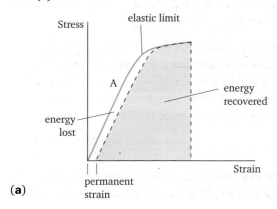

**(a)**

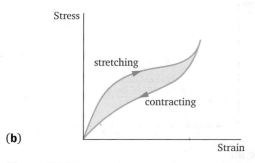

**(b)**

**Figure 4.84**

The work done in stretching is the area under the graph line labelled **A**. When the force is removed, only the energy that corresponds to the shaded area in Figure 4.84(a) is recovered.

For rubber, there may be no permanent strain when the stretching force is removed but energy is still lost. The rubber gains internal energy as it is stretched and this is used to heat the surroundings so is not retrievable. Figure 4.84(b) shows a typical graph obtained when rubber is first stretched and then allowed to contract by gradually reducing the tensile force.

The shaded area in Figure 4.84(b) is the energy that is not regained when the force is reduced.

## Measurement of Young's Modulus

A suitable apparatus for investigating the behaviour of a wire under stress and for measuring Young's modulus is shown in Figure 4.85.

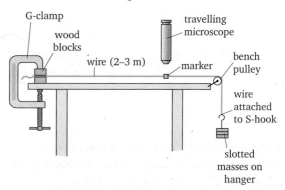

**Figure 4.85**

### Selecting apparatus

In order to reduce uncertainties in the measurement of the extension, the extension should be as large as possible. This is achieved by using long lengths of thin wire. A small diameter also requires lower loads to produce measurable extensions than thicker wires. However, use of very thin wires increases the uncertainty in the measurement of the diameter and hence in the area of cross-section.

### Procedure and analysis

- Sufficient mass is suspended from the wire to remove any kinks and to make the wire taut. This mass is ignored in future measurements of the stretching force.
- The initial length $l$ of the wire, between the fixed end and a marker (a small piece of masking tape), is measured using one or more metre rulers.
- A mass $m$ is added to the wire to provide the stretching force $mg$.
- Wait for a short time to ensure that the wire does not continue to extend. If it does you have added too much mass to the wire and may need to start again! (See yielding on page 59.)
- The extension $\Delta l$ produced by this force is measured ($\Delta l$ = new length $- l$). To make an accurate measurement of $\Delta l$, vernier callipers or a travelling microscope may be used.
- This is repeated until you have at least five sets of values for extending force and extension. (You will probably have many more sets in practice.)
- Plot a graph of $F$ against $\Delta l$.
- The gradient of this graph is $\dfrac{F}{\Delta l}$.
- $E$ is then the gradient $\times \dfrac{l}{A}$.

- *A* is determined by measuring the diameter *d* at several places, using a micrometer, and taking an average. (Take care not to squash the wire while making the measurement.)

$$A = \frac{\pi d^2}{4}$$

If you prefer, you could plot a stress–strain graph instead of a force–extension graph. The graphs have the same general shape. The gradient of the stress–strain graph gives *E* directly.

## Problem of Changing Area

When stretching metal wires within the elastic limit, it is reasonable to assume that the area of cross-section remains unchanged during the experiment.

In practice, as the length increases the area of cross-section falls, the volume of wire being approximately constant. The percentage fall in area is the same as the strain produced. For a metal wire, the maximum strain is about 0.01 for the linear region, which is a change of only 1%. The effect of the assumption is therefore small compared with uncertainties in measurement such as that of the diameter of the wire.

The strains when investigating rubber are much larger, so account has to be taken of the change in area when making measurements to produce a stress–strain graph.

Values for Young's modulus, yield stress and breaking stress for some materials are given below.

**Table 4.5**

|  | *E*/Pa | Yield stress/MPa | Breaking stress/MPa |
| --- | --- | --- | --- |
| Aluminium | $7.1 \times 10^{10}$ | 50 | 80 |
| Copper | $1.2 \times 10^{10}$ | 75 | 150 |
| Steel | $2.1 \times 10^{11}$ |  | 3000 |
| Brass | $1.0 \times 10^{11}$ | 450 | 550 |
| Perspex | $3.0 \times 10^{9}$ |  | 50 |
| Rubber | $2.0 \times 10^{7}$ |  | 17 |

### Test your understanding

1  A force of $1.5 \times 10^6$ N is applied to a steel bolt of diameter $9.0 \times 10^{-2}$ m. Calculate:
   (a) the strain in the bolt
   (b) the energy stored per unit volume in the bolt.
2  Calculate the force required to produce a strain of $5.0 \times 10^{-4}$ in an aluminium wire of diameter 1.8 mm.

3  A rod in a suspension bridge has a length 55 m and cross-sectional area $5.0 \times 10^{-3}$ m². It supports a load of 5000 kg. Young's modulus of the steel in the rod is $2.0 \times 10^{11}$ Pa. The rod operates within the Hooke's law region. Calculate:
   (a) the stress in the rod
   (b) the strain in the rod
   (c) the elastic strain energy stored in the rod.
4  The ultimate breaking stress for the material used to form a crown for a tooth is 180 MPa. It breaks when the strain is 2.5%. Assuming that it obeys Hooke's law until it breaks calculate:
   (a) Young's modulus of the material
   (b) the energy per unit volume needed to break the crown.
5  Figure 4.86 shows the force–extension graph for a rubber cord.

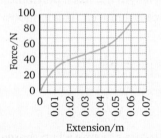

**Figure 4.86**

Determine the energy used to stretch the cord by 0.40 m.
6  Figure 4.87 shows the variation of stress with strain for a sample of copper.

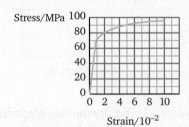

**Figure 4.87**

   (a) Use this data to determine the Young modulus of copper.
   (b) Determine the work done per unit volume in extending the wire:
      (i)   to the limit of proportionality
      (ii)  to a strain of 0.1.

# Capacitance and Exponential Decay

## Capacitors

The simplest capacitor is a parallel plate capacitor. This is made from two parallel conducting plates separated by an insulator, as shown in Figure 4.88. The insulator may be air or a sheet of a material such as polythene. The symbol for a capacitor is also shown in Figure 4.88.

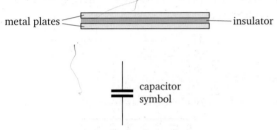

**Figure 4.88**

The capacitor stores energy when charge is transferred from one of its plates to the other as shown in Figure 4.89.

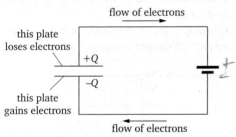

**Figure 4.89**

When charge is transferred (by moving electrons) the potential difference between the plates increases. We then say that the capacitor has been **charged**.

Note that it is common to refer to the process as a charge being placed on the capacitor although the process is one of transfer of charge. It is also common to say that the capacitor is charged to a certain potential difference.

## Capacitance

The capacitance of a capacitor is the charge that has to be placed on the capacitor to change the potential difference (p.d.) between the plates by 1 volt. It is the charge per volt on the capacitor.

It follows that when a charge $Q$ produces a potential difference $V$ between the plates the capacitance $C$ is given by:

$$C = \frac{Q}{V}$$

The unit of capacitance is the **farad F**. One farad is one coulomb per volt, $C\,V^{-1}$.

You may find it easier to remember the relation between $Q$, $V$ and $C$ in the form:

$$Q = VC$$

You need to be alert in problems so as not to confuse the unit C with the symbol for capacitance $C$. The unit of:

$C$ is F (farad)
$Q$ is C (coulomb)
$V$ is V (volt).

EXAMPLE: When there is a charge of $5.0 \times 10^{-3}$ C (5.0 mC) on a capacitor, the p.d. between the plates is 3.0 V. Calculate the capacitance of the capacitor.

$$C = \frac{Q}{V}$$

$$C = \frac{5.0 \times 10^{-3}}{3.0} = 1.6 \times 10^{-3}\,F$$

This would usually be given as 1.6 mF or 1600 μF.

### Graph of $Q$ against $V$

The graph in Figure 4.90 shows how the charge $Q$ on the capacitor varies as the potential difference $V$ between the plates of a capacitor changes.

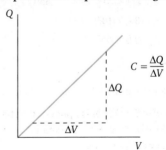

**Figure 4.90**

The slope of this graph gives the value of the capacitance.

## Markings on a Capacitor

There are two important values that you need to be aware of when choosing a capacitor. These are given in catalogues for all capacitors and are marked on capacitors that have large enough dimensions.

The markings give:
- the capacitance in F
- the working voltage, which is the maximum safe voltage that can be used between the plates of the capacitor.

For example, an electrolytic capacitor with markings 2200 μF, 15 V signifies that:
- a charge of 2200μC will produce a potential difference of 1 V between the plates of the capacitor
- the p.d. between the plates must not exceed 15 V.

If the maximum p.d. is exceeded, charge may leak through the insulating layer so that charge can no longer build up. It is possible for the insulation to break down completely, which means that the capacitor is destroyed.

Electrolytic capacitors also carry symbols (+ and/or −) to show which terminal should be the more positive when connected in a circuit. The correct connection in a charging circuit is shown in Figure 4.91.

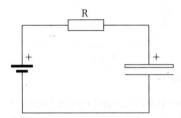

**Figure 4.91**

The insulating layer in an electrolytic capacitor is easily destroyed when connected the wrong way round. In extreme cases, an incorrect connection is also hazardous since a build up of gases inside the capacitor can cause it to explode.

## Structure of Capacitors

Capacitors may be made from sheets of metal foil with wax-impregnated paper between the plates. The arrangement is rolled up into a cylinder as shown in Figure 4.92.

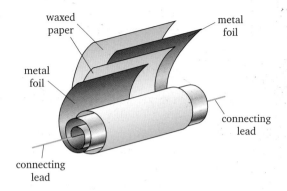

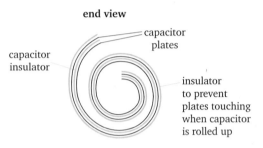

**Figure 4.92**

The cylinder is enclosed in an insulated wrapping with wires to connect the metal sheets to the outside world. Large values of such capacitors have a large dimension. However, they can be designed to work effectively with a high potential difference between the plates.

The structure of an electrolytic capacitor is shown in Figure 4.93.

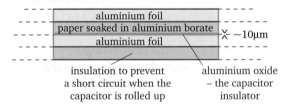

**Figure 4.93**

The insulating layer is formed by electrolysis. This layer is very thin so that capacitors with a large capacitance can be made with small physical dimensions. However, these capacitors can only be used with relatively low voltages.

## Capacitors in Parallel

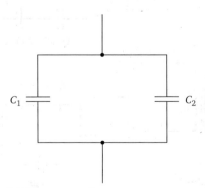

**Figure 4.94**

Figure 4.94 shows two capacitors connected in parallel. The effective capacitance $C$ of the combination is given by:

$$C = C_1 + C_2$$

- The value is always larger than that of the highest capacitance used in the combination.
- The capacitor with the lowest maximum safe working voltage gives the maximum safe voltage across the combination.
- The potential difference across each capacitor in the combination is the same.

## Capacitors in Series

**Figure 4.95**

Figure 4.95 shows two capacitors connected in series. The effective capacitance $C$ of the combination is given by:

$$\frac{1}{C} = \frac{1}{C_1} + \frac{1}{C_2}$$

- The capacitance of the combination is always smaller than that of the smallest capacitance used in the combination.
- When the capacitors in the combination are the same, the maximum safe voltage across the combination is doubled.
- Both capacitors have the same charge.
- The largest potential difference is across the capacitor with the lower capacitance.

### Test your understanding

1 Calculate the voltage across a 470 μF capacitor that has a charge of 1.2 mC.
2 Calculate the combined capacitance of the combinations in Figure 4.96.

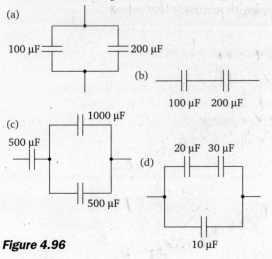

**Figure 4.96**

3 Calculate the value of the capacitor that you would place:
  (a) in parallel with a 100 μF capacitor to produce a capacitance of 320 μF
  (b) in series with a 100 μF capacitor to produce a capacitance of 50 μF.
4 Calculate the total charge on two 100 μF capacitors when connected in parallel to a 3.0 V supply.
5 Calculate the charge on each capacitor in the circuit shown in Figure 4.97.

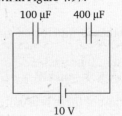

**Figure 4.97**

## Energy Stored in a Capacitor

Figure 4.98 shows the variation of voltage across a load resistor with charge flow when the voltage remains constant (as when a battery discharges through a lamp).

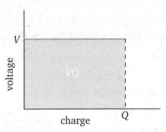

**Figure 4.98**

The total energy delivered by the battery when a charge $Q$ flows is $VQ$. Notice that this is the area under the V–Q graph.

When a capacitor discharges through a resistor, the potential difference across the capacitor and the resistor falls. Figure 4.99 shows how voltage varies with charge for a capacitor.

(This is similar to Figure 4.90.)

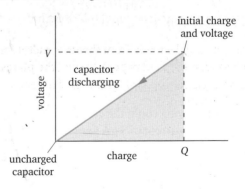

**Figure 4.99**

The area under the $V$-$Q$ graph again gives the total energy that can be delivered by the capacitor. This is the energy that was stored in the capacitor and is the area of the shaded triangle in Figure 4.99.

For a capacitor with an initial charge $Q$ and potential difference $V$, the energy stored is given by:

$$W = \tfrac{1}{2} QV$$

This can be thought of as being the average potential difference during the discharge multiplied by the charge.

Since $Q = VC$, other formulae that give the energy stored are:

$$W = \tfrac{1}{2} \frac{Q^2}{C} \text{ and } W = \tfrac{1}{2} CV^2$$

In practice, the last of these is usually the most useful since capacitance values and the potential difference are usually the known quantities.

## Use of capacitors for energy storage

Capacitors are used for storing energy in an electronic flash unit for use in photography. The energy stored is discharged very rapidly producing a high power light for a short time.

The potential difference required for electronics using microcircuits is about 5 V. Electrolytic capacitors capable of handling this potential difference can be manufactured having a very large capacitance and taking up a relatively small space. They are used to provide back-up supplies.

EXAMPLE ONE: The flash in a camera flash unit has a capacitance of 2200 μF. It is charged to 9.0 V. The capacitor is discharged in 30 μs to produce the flash. Calculate the average power delivered by the capacitor.

Energy stored $W = \tfrac{1}{2} CV^2 \quad = 2200 \times 10^{-6} \times 9.0^2$
$$= 0.089 \text{ J}$$

Average power $= \dfrac{W}{t}$

$$= 2970 \text{ W}$$

EXAMPLE TWO: A capacitor of 0.02 F is charged to 5 V.

(a) Calculate the charge on the capacitor.

(b) Calculate the energy stored by the capacitor.

(c) For how long could the charged capacitor supply a current of 5 μA to maintain the information stored in the memory of a computer?

(a) Charge $Q = VC = 5 \times 0.02 = 0.1 \text{C}$

(b) Energy stored $= \tfrac{1}{2} CV^2 = \tfrac{1}{2} \times 0.02 \times 5^2$
$$= 0.25 \text{ J}$$

(c) $Q = It$, so $t = Q/I = 0.1/(5 \times 10^{-6})$
$$= 20\,000 \text{ s or } 5.6 \text{ h}$$

### Test your understanding

1 Complete the following table.

**Table 4.6**

| Capacitance | p.d. | Charge | Energy stored |
|---|---|---|---|
| 3.2 μF | 12 V | | |
| | 3.5 V | 12 mC | |
| | | 100 μC | 1.2 mJ |
| | 5.0 V | | 0.10 mJ |

2 A 2200 μF capacitor is charged to a potential difference of 6.0 V in 3.2 s. Calculate:
  (a) the average charging current
  (b) the energy stored by the capacitor.
3 The graph of voltage $V$ against charge $Q$ for a capacitor is shown in Figure 4.100.

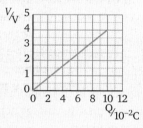

**Figure 4.100**

Calculate:
  (a) the value of the capacitance
  (b) the energy stored when the potential difference is 3.5 V.
4 Calculate the average power delivered by a 20 000 μF capacitor that is initially charged to 25 V and discharges to a negligible voltage in 0.15 ms.

## Factors Affecting Capacitance

The capacitance of a parallel plate capacitor is determined by:
- the surface area $A$ of one of the plates
- the separation $d$ of the two plates
- the material between the plates, often referred to as the **dielectric**.

The formula for capacitance is:

$$C = \frac{\varepsilon A}{d} \text{ or } C = \frac{\varepsilon_0 \varepsilon_r A}{d}$$

where:

- $\varepsilon$ is a constant for the material called its **permittivity**
- free space (a vacuum) has the lowest possible permittivity $\varepsilon_0$ (called the **permittivity of free space**)
- $\varepsilon_0$ has a value $8.9 \times 10^{-12}\,\mathrm{F\,m^{-1}}$
- for other materials data books give the **relative permittivity** $\varepsilon_r$. The relative permittivity of polythene is approximately 2
- $\varepsilon = \varepsilon_0 \varepsilon_r$
- the permittivity of air can be assumed to be the same as that of free space since its relative permittivity is 1.0005 (1 to 0.005% precision).

(Permittivity is also involved in the work on electric fields that you will study in Module 5.)

## Measuring Capacitance

There are now available digital meters that can measure capacitance directly. These are useful when performing experiments to confirm the formulae for capacitors in series and in parallel.

However, the capacitance of a simple parallel plate capacitor is too small for such instruments to measure. Small capacitance can be measured by using an electrometer to measure the charge for a given supply voltage. A high voltage is needed to produce a large enough charge to be measurable.

### Reed switch

Use of a reed switch provides a convenient and instructive method of measuring small capacitance.

Figure 4.101 shows a simple circuit that can be used to charge and discharge a capacitor.

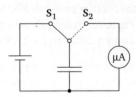

**Figure 4.101**

When the switch is in the position $S_1$ the capacitor charges to the supply voltage. When in position $S_2$ the capacitor discharges through the meter. A deflection shows that a charge is flowing through the meter.

Figure 4.102 shows a similar circuit containing a reed switch.

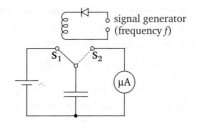

**Figure 4.102**

In a reed switch, the switch vibrates rapidly between $S_1$ and $S_2$. The vibration is caused by an electromagnet connected to an alternating voltage from a signal generator. The switching takes place at the oscillator frequency $f$ (usually about 300 Hz). Using this circuit, the capacitor charges to the supply voltage and discharges through the meter $f$ times each second.

For each oscillation of the switch the charge that flows through the meter is given by:

$$Q = VC$$

Since there are $f$ oscillations each second, the charge per second through the meter is $fCV$.

Charge flow per second is the current $I$ recorded by the meter so:

$$I = fVC$$

The capacitance $C$ of the capacitor is therefore given by the equation:

$$C = \frac{I}{fV}$$

With a suitable method for measuring a small capacitance, the factors that affect the capacitance can be investigated.

### Testing the equation $C = \dfrac{\varepsilon_0 \varepsilon_r A}{d}$

The capacitor used in experiments usually consists of two flat square plates (0.25 m × 0.25 m). The plates can either be separated by an insulator, such as polythene sheet one millimetre or so thick, or small pieces of polythene placed at the corners so that the plates are separated by air.

### Experiment to show that $C \propto 1/d$

- The area of overlap of the plates is kept constant. Using the full area of the plates will give the maximum capacitance and reduce uncertainties in its measurement.
- The plates are separated by small pieces of polythene or glass cut from a microscope slide.
- The capacitance is measured for five different separations, the separation being adjusted by piling up separators.

- The thickness of one separator is measured using a micrometer.
- A graph of $C$ against $1/d$ should be a straight line as shown in Figure 4.103.

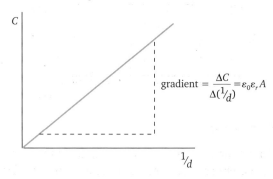

$$\text{gradient} = \frac{\Delta C}{\Delta(1/d)} = \varepsilon_0\varepsilon_r A$$

**Figure 4.103**

The gradient of this graph is $\varepsilon_0\varepsilon_r A$. Since $A$ can be determined and since for an air gap $\varepsilon_r = 1$, this experiment also enables the permittivity of free space $\varepsilon_0$ to be found.

### Experiment to show that $C \propto A$

- The separation of the plates is kept constant using spacers. The separation should be as small as possible to produce the largest possible capacitance and reduce uncertainties in its measurement.
- The capacitance is measured for five different areas of overlap of the plates.
- Overlapping areas between half and the maximum area are suitable.
- The area is determined from the dimensions of the overlapping areas.
- A graph of $C$ against $A$ should be a straight line as shown in Figure 4.104.

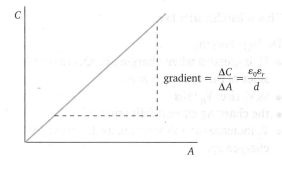

$$\text{gradient} = \frac{\Delta C}{\Delta A} = \frac{\varepsilon_0\varepsilon_r}{d}$$

**Figure 4.104**

### Measurement of relative permittivity

Measuring the capacitance first with a sheet of polythene and then using small spacers of the same thickness, the relative permittivity of polythene can be determined.

**Test your understanding**

1 State three ways of increasing the capacitance of a capacitor.
2 A capacitor is made from two overlapping plates, each of area $2.5 \times 10^{-3}\,\text{m}^2$.
   Calculate the capacitance:
   (a) when the plates are separated by 0.15 mm, with air between the plates
   (b) when the plates are separated by 0.05 mm and the space between the plates is filled with a material of relative permittivity 2.2.
3 (a) Calculate the separation of the plates of a capacitance of a 1 pF capacitor made from two plates each of surface area $2.2 \times 10^{-4}\,\text{m}^2$ when the material between the plates is air.
   (b) A capacitor of the same capacitance is manufactured on a microchip. The area of each 'plate' is now $1 \times 10^{-6}\,\text{m}^2$. The material between the plates has a relative permittivity of 1.2. Calculate the separation of the plates in this case.
4 In Figure 4.105, the 200 pF capacitor is being charged to a voltage of 6.0 V and then discharged completely at a frequency of 300 Hz.

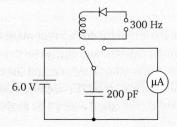

**Figure 4.105**

   (a) Calculate the discharge current.
   (b) What would be the effect on the current of halving the frequency of the supply to the reed switch and halving the supply voltage?
5 (a) Draw accurately a graph showing the variation of capacitance with separation for plates of area $0.063\,\text{m}^2$ separated by air for separations up to 5.0 mm.
   (b) Draw on the same axes a graph showing the effect of halving the area of the plates.

# Charging a Capacitor

The circuit shown in Figure 4.106 can be used to show how the potential difference across a capacitor varies with time when it is charged through a resistor.

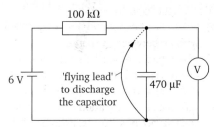

**Figure 4.106**

It is important that the voltmeter is a high resistance digital meter so that negligible charge flows through it when charging commences.

As an alternative to manual measurement, this type of experiment is ideally suited to data capture using a voltage sensor and a computer. The computer can make measurements at very short time intervals so that circuits can be investigated in which charging is much faster.

To make measurements during charging, the capacitor is first discharged using the flying lead. When the flying lead is disconnected the capacitor will commence charging. The voltage $V_C$ across the capacitor is measured every 10 s as the capacitor charges.

The graph labelled $V_C$ in Figure 4.107 shows how the p.d. across the capacitor varies with time $t$.

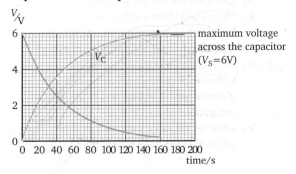

**Figure 4.107**

A lower resistor will increase the rate at which the capacitor charges.

Since $Q \propto V$, a graph of $Q$, the charge on the capacitor, against time $t$ has the same shape as the $V$-$t$ graph, the final charge being $2800\,\mu C$, as shown in Figure 4.108.

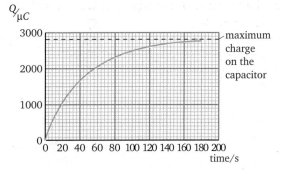

**Figure 4.108**

When a similar experiment is conducted with the voltmeter connected across the resistor to measure $V_R$, the result is the graph labelled $V_R$, shown in Figure 4.107.

Since the current $I$ in the circuit is proportional to the p.d. across the resistor, a graph of discharge current against time has a similar shape to the graph of $V_R$ against time with the initial current being $60\,\mu A$ as shown in Figure 4.109.

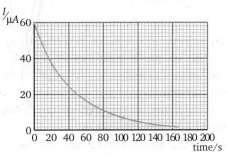

**Figure 4.109**

Note that at all times:

$$V_S = V_C + V_R$$

This is **Kirchhoff's law**.

During charging:
- $V_R$ is greatest when charging begins (at $t = 0$) since at this time $V_C$ is zero
- as $V_C$ rises $V_R$ falls
- the charging current falls (since $V_R \propto I$)
- $V_C$ increases at a slower rate as the capacitor charges up.

## Discharging a Capacitor

The circuit in Figure 4.110 can be used to show how the voltage across a capacitor varies with time when it is discharged through a resistor.

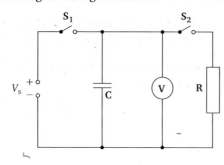

**Figure 4.110**

- When $S_1$ is closed and $S_2$ open, the capacitor charges.
- When $S_1$ is opened and $S_2$ closed, there is a voltage across the resistor so there is a current through the resistor ($I = V/R$).
- Charge flows in the circuit so the voltage across the capacitor falls.
- This leads to a fall in discharge current and the capacitor discharges more slowly.
- The voltage will take a very long time indeed to reach zero (the mathematical treatment suggests it never will).

Figure 4.111 shows the graph of voltage $V_C$ (or $V_R$) against time $t$ for the discharge.

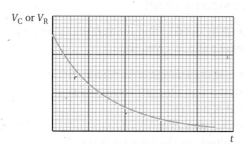

**Figure 4.111**

Note that, when discharging, the voltage across the capacitor must always be equal to that across the resistor.

The rate at which charge flows off the capacitor is given by:

$$I = \frac{\Delta Q}{\Delta t} = -\frac{Q}{RC}$$

since:

$$V = \frac{Q}{C}$$

The minus sign is the mathematical way of saying that the charge is decreasing with time.

Since $V \propto Q$ and $Q \propto I$, the same shape as for the graph shown in Figure 4.111 is obtained for plots of:

- discharge current $I$ against $t$
- charge on the capacitor $Q$ against $t$.

These graphs are shown in Figure 4.112 and Figure 4.113.

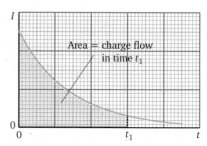

**Figure 4.112**

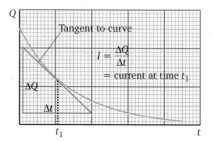

**Figure 4.113**

The following relationships between these graphs should be noted.

- The slope of the $Q$-$t$ graph at any time gives the discharge current at that time.
- The area under the current–time graph between two given times is the charge that has flowed during that time.

The total area under the curve in Figure 4.112 effectively gives the initial charge on the capacitor.

The gradient at $t = 0$ in Figure 4.113 is the initial discharge current.

## Exponential Curves

Whenever the rate at which a quantity decreases is proportional to how much of it there is at a given time, the decay is **exponential**. The curves in Figures 4.111 to 4.113 are all **exponential curves**.

The equation that shows how charge varies with time is:

$$Q = Q_0 e^{-t/RC}$$

where $Q_0$ is the initial charge on the capacitor, i.e the charge at $t = 0$.

Equations for the variation of p.d. $V$ across the capacitor and the discharge current $I$ are similar.

$$V = V_0 e^{-t/RC} \text{ and } I = I_0 e^{-t/RC}$$

e is a constant with a value of 2.72 (to 3 significant figures). Like $\pi$, this is a 'never-ending' number.

The product $RC$ is the **time constant** of the circuit. It is this that determines the rate of discharge (or charge). When $R$ is in $\Omega$ and $C$ is in F the time constant is in seconds. It is essential that you convert to these units when doing calculations.

### Properties of an exponential curve

All exponential decay curves have the same properties. The following are the properties of the $Q$-$t$ graph in Figure 4.114.

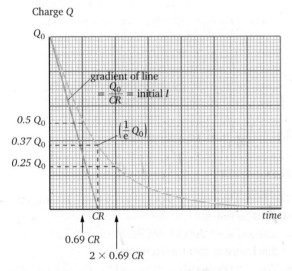

**Figure 4.114**

- $Q$ always falls by the same factor in the same time interval
- $Q$ falls to $1/e$ (0.37) of its initial value in a time equal to the time constant $RC$.
  When the initial charge is $Q_0$,
  after $RC$ seconds $Q = 0.37\,Q_0$,
  after $2RC$ seconds $Q = 0.37 \times 0.37\,Q_0$
  $= 0.14\,Q_0$, etc.
- After 5 time constants the charge remaining is 0.0069 (0.69%) of the initial charge, and for practical purposes the capacitor can be assumed to have completely discharged.
- The time taken to halve $T_{\frac{1}{2}}$ is always the same and equal to $0.69RC$.

$$T_{\frac{1}{2}} = 0.69RC$$

after $T_{\frac{1}{2}}$ seconds the charge is $\frac{1}{2}Q_0$

after $2T_{\frac{1}{2}}$ the charge is $\frac{1}{2} \times \frac{1}{2}Q_0 = 0.25Q_0$, etc.

- The rate of change of charge (the initial discharge current) is the initial gradient $\dfrac{Q_0}{RC}$.

## Using the Decay Curve

It is often necessary to determine the time constant or other quantities from the decay curve. Take care not to confuse time constant and the time to halve.

The easiest approach is to determine the time to halve and then use the equation $T_{\frac{1}{2}} = 0.69RC$.

When you know the time constant, $R$ or $C$ can be found if one of them is known.

## To Test for an Exponential Decay

### Non-graphical approach

- Read the value at any time.
- Calculate half that value.
- Determine from the graph the time taken to halve.
- Repeat for other starting values.
- If the graph is exponential the time should always be the same.

### Graphical approach

Alternatively, a graph of the log or ln of the value (e.g. log $Q$ or ln $Q$) against time should produce a straight line of negative gradient (see page 16).

### Using the formula

You need to be able to do two types of calculation for exponential decay.

**1** You may need to find the voltage, charge or current after a given time.

To find the voltage after a given time:
- determine the time constant $RC$
- calculate $t/RC$
- calculate $e^{-t/RC}$
- multiply by the initial voltage $V_0$.

**2** You may need to find how long it takes to reach a given voltage, charge or current.

To find the time to reach $V$ when the initial voltage is $V_0$:
- determine $V/V_0$
- calculate ln $(V/V_0)$
- multiply by $-1$
- multiply by $RC$.

### Using a spreadsheet

Predicting the discharge curve for a capacitor is an ideal way to learn the process of step and repeat calculations using a spreadsheet. This process is a useful tool. It is not a requirement for examinations, but it is included in **Appendix B** (see page 136) for interested students.

**Test your understanding**

1  Figure 4.115 shows the voltage across a capacitor at a given time when being charged.

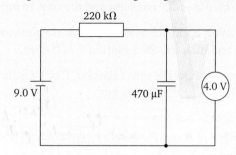

**Figure 4.115**

For this time calculate:
(a) the voltage across the resistor
(b) the charging current
(c) the charge on the capacitor.

2  Complete the following table showing correct corresponding values when a capacitor discharges through a resistor.
In the table, capacitance = $C$, resistance = $R$, time constant = $T$ and time for the voltage to halve = $T_{\frac{1}{2}}$.

**Table 4.7**

| C | R | T | $T_{\frac{1}{2}}$ |
|---|---|---|---|
| 2200 μF | 100 kΩ | | |
| 470 μF | | 5.0 s | |
| | 2.2 kΩ | | 2.0 ms |
| 100 nF | | | 4.5 ms |

3  The graph in Figure 4.116 is the current–time graph for a capacitor discharging through a 22 kΩ resistor.

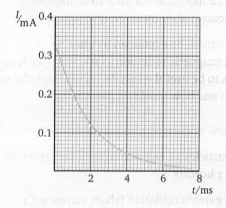

**Figure 4.116**

(a) Determine the time constant of the circuit.
(b) Calculate the capacitance of the capacitor in the circuit.
(c) Use the graph to estimate the charge on the capacitor before discharge commenced.

4  A 4.7 μF capacitor is charged to 10 V and then discharged through a 470 kΩ resistor. Calculate the time taken for the potential difference across the capacitor to fall:
(a) to $\dfrac{1}{e}$ of its initial value
(b) to 5.0 V
(c) to 2.5 V
(d) to 4.0 V.

5  For the discharge circuit in question 5, calculate:
(a) the potential difference after 1.0 s
(b) the charge remaining on the capacitor after 0.50 s.

6  Figure 4.117 shows the charge–time graph for a 500 μF capacitor when discharging through a resistor.

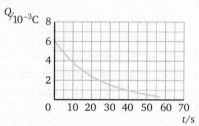

**Figure 4.117**

(a) Calculate the resistance of the resistor.
(b) Use the graph to determine the initial discharge current (recall that $I = \dfrac{\Delta Q}{\Delta t}$).
(c) Calculate the potential difference to which the capacitor was charged.

# Quantum Phenomena

## Waves and Particles

In the course so far we have considered that energy transfer takes place in one of two ways.

- A particle (having mass) is given kinetic energy. This particle delivers all or part of this energy when it interacts with other particles during transit or on arrival at its destination;
- By wave motion in which one part of a medium (e.g. an atom or molecule) is given energy and this passes the energy onto the next part of the medium, and so on.

One of these processes, on its own, can be used to explain the effects observed when energy transfer takes place on a large scale. For example, the first process is sufficient to explain energy transfer when throwing a ball and the second to explain energy transfer in sound.

However, when transferring energy in the form of electromagnetic waves or using small particles such as electrons, effects are observed that cannot be explained using one of theses processes on its own.

**The wave theory is needed to explain where the energy goes.**

**The particle theory is needed to explain what happens when the energy is detected.**

We say that electromagnetic waves and electrons exhibit **wave–particle duality**.

Note that wave–particle duality extends to all matter, but the wave effects are too small to observe on a large scale.

### Wave properties

- Wave energy spreads out from the source, becoming weaker and weaker (less energy per second per square metre) as distance from the source increases.
- Waves can diffract around obstacles so that the energy spreads out.
- Two or more sources can produce waves that superpose to produce interference patterns with regions of maximum and minimum intensity, showing that energy can only travel in certain directions.

### Particle properties

- A particle has mass.
- When moving, a particle has energy ($\frac{1}{2} mv^2$) and momentum ($mv$).
- A particle obeys Newton's laws of motion.
- A particle will travel in a straight line unless a force acts on it.
- A particle will have a constant energy until it interacts.
- A particle can give up all or part of its energy in a single interaction.

You will need to keep these properties in mind as you study this section.

## Electromagnetic Radiation

### Wave nature of electromagnetic radiation

Electromagnetic radiation:
- can be diffracted by a narrow slit
- can produce interference patterns using a two-slit arrangement
- always has the same speed in a given medium whatever its intensity
- spreads out from a point source, becoming weaker and weaker (less energy per second per square metre) as distance from the source increases, obeying an inverse square law.

These can all be explained by treating electromagnetic radiation as a wave. The theory proves to be useful when deciding where the energy will go and how quickly it gets there.

### Particle nature of electromagnetic radiation

The 'particles' of electromagnetic radiation are called **photons**.

When gamma radiation (short wavelength electromagnetic radiation) is detected using a **Geiger–Müller** (G–M) tube, the scalar records random counts. This suggests that energy is detected in random pulses rather than continuously, as you would expect for a wave.

Each pulse is produced by a **photon** of electromagnetic radiation producing ionisation of the gas in the G–M tube.

- A photon is a wave-packet that contains a certain amount of energy. It is a quantum of electromagnetic radiation.
- For a given frequency of radiation, photons all have the same energy.
- For a given frequency, energy can only be delivered in multiples of the photon energy.

The energy $E$ of a photon is given by:

$$E = hf$$

where $f$ is the frequency of the radiation and $h$ is Planck's constant ($6.6 \times 10^{-34}$ J s).

$$\text{Since } f = \frac{c}{\lambda}, \qquad E = \frac{hc}{\lambda}$$

EXAMPLE: Green light has a wavelength of $5.0 \times 10^{-7}$ m. Determine the frequency and photon energy of green light.

The frequency of green light is:

$$\frac{3.0 \times 10^8}{5.0 \times 10^{-7}} = 6.0 \times 10^{14} \text{ Hz}$$

The photon energy of green light is:

$$6.0 \times 10^{14} \times 6.6 \times 10^{-34} = 4.0 \times 10^{-19} \text{ J}$$

## Intensity of Radiation

Intensity of electromagnetic radiation is the total energy arriving per second per square metre.

Assuming the energy of electromagnetic radiation to arrive as a wave:
- the amplitude of the wave is related to the intensity (intensity $\propto$ amplitude$^2$)
- the energy will arrive evenly spread over a surface
- all parts of the surface will be receiving energy at all times.

Assuming the energy of an electromagnetic wave to be a stream of particles:
- at a given time some parts of the surface will be receiving energy but others will not
- the intensity is the sum of the energy of all the photons that is delivered to each square metre of a surface per second
- if the intensity doubles there are twice as many photons striking a given area of surface each second.

EXAMPLE: How many photons of red light of frequency $4.8 \times 10^{14}$ Hz are arriving each second on each square millimetre of a surface when the intensity of the light is 5.0 mW m$^{-2}$?

Photon energy $= 3.17 \times 10^{-19}$ J

Energy per s per mm$^2$

$$= 5.0 \times 10^{-3} \times 10^{-6} = 5.0 \times 10^{-3} \text{ J}$$

Number of photons arriving per s on 1 mm$^2$

$$= \frac{5.0 \times 10^{-3}}{3.17 \times 10^{-19}} = 1.6 \times 10^{16}$$

## The Photoelectric Effect

The photoelectric effect is the emission of electrons from a surface when electromagnetic radiation is incident on it.

The observations made, when the photoelectric effect occurs, can only be explained by assuming that energy is carried by photons.

You need to recall one experiment that demonstrates this effect. Two are described below.

**Experiment 1**

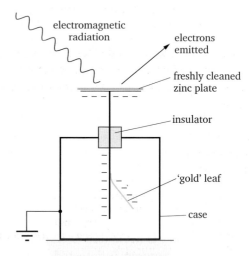

**Figure 4.118**

The arrangement is shown in Figure 4.118. The electroscope shows a deflection of the gold leaf when charge is placed on the central electrode. The leaf hangs at an angle, as shown, until the charge is removed.

A clean zinc plate is placed on the central electrode.

The electroscope is charged negatively by connecting the negative terminal of an EHT supply to the central electrode and the positive of the supply to the case of the electroscope.

The zinc plate is illuminated with radiation. When ultraviolet radiation is used the electroscope discharges, suggesting that negative charge is being removed from the plate.

When visible radiation is used no discharge occurs. There is a **threshold frequency** below which no electrons are released.

The discharge occurs **immediately** even when using very low intensity ultraviolet radiation, whereas no discharge occurs even with very intense visible radiation.

When the plate is positively charged no discharge occurs when the plate is illuminated.

## Conclusions

If the energy were arriving as a wave then:
- with low intensity radiation a delay would be expected before electron emission
- there should be many electrons emitted by high intensity radiation whatever its frequency.

The observations can only be explained by assuming that the energy is arriving in well-defined quantities in the form of photons.

The existence of a threshold frequency suggests that the energy of a photon depends on its frequency. Low energy photons do not have sufficient energy to liberate an electron.

Electrons can only be emitted when it receives sufficient energy instantaneously. An electron cannot save up energy until it has enough to leave the surface.

**Experiment 2**

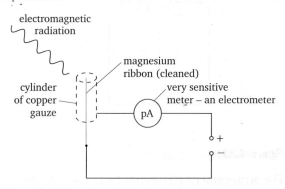

**Figure 4.119**

Using the apparatus in Figure 4.119, electrons liberated from the central electrode enable a small current to flow in the circuit, which can be measured using the electrometer.

The electrode is illuminated by radiation of different frequencies.

There is a current:
- only when light of frequency greater than the threshold frequency is incident on the surface
- only when the outer electrode is more positive than the central electrode.

Provided that the frequency is above a threshold frequency:
- there is a current even when the intensity is low
- there is no noticeable time delay before a current starts even when the intensity is low
- increasing the intensity increases the current.

Increasing the intensity of radiation that has a frequency lower than the threshold frequency does not produce a current.

The conclusions drawn from these observations are the same as those in the previous experiment.

## Work Function

Electrons are bound to the surface of a metal. This means that energy is needed to liberate them.

The **work function** $\Phi$ is the **minimum** energy required to liberate an electron from a surface.

It is the minimum energy because electrons that are not near to a surface will need more energy to liberate them.

This can be represented by the energy level diagram shown in Figure 4.120. No electron can exist with energy between $-\Phi$ and 0.

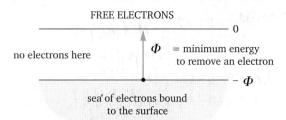

**Figure 4.120**

An electron that is just free has zero energy. Electrons within a metal have energies equal to or less than $-\Phi$.

An electron with an energy of $-\Phi$ has to be given an energy of $\Phi$ for it to just become free.

Work functions are usually given in electron-volt (eV).

$$1 \text{ eV} = 1.6 \times 10^{-19} \text{ J}$$

Table 4.8 shows the work function $\Phi$ of some metal surfaces in eV and J.

**Table 4.8**

| Metal | $\Phi$/eV | $\Phi$/J |
|---|---|---|
| Potassium | 2.2 | $3.5 \times 10^{-19}$ |
| Zinc | 4.3 | $6.9 \times 10^{-19}$ |
| Magnesium | 3.6 | $5.8 \times 10^{-19}$ |
| Tungsten | 4.6 | $7.4 \times 10^{-19}$ |

Note that contamination (e.g. oxidation) will increase the work function of a surface.

## The Photoelectric Equation

When a photon of energy higher than the threshold energy is incident on a surface all its energy is given to an electron.

The photon energy ($hf$) can be used to:
- provide the electron with energy to leave the surface
- provide the electron with kinetic energy.

This is summarised in the conservation of energy equation:

$$hf = \Phi + E_{k(max)}$$

photon energy = work function + maximum electron KE

Note that the equation gives the maximum KE. For those electrons that are more tightly bound, more energy is used to liberate them so there is less to provide kinetic energy.

Figure 4.121 is a graph showing how the maximum kinetic energy $E_{k(max)}$ emitted by the electrons varies with photon energy $hf$. The intercept on the photon energy axis is the work function of the surface. A graph of $E_{k(max)}$ against $f$ would have the same shape.

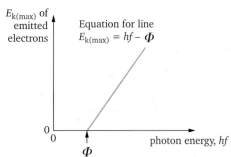

**Figure 4.121**

## Practical investigation of the photoelectric equation

The equation can be investigated using the system shown in Figure 4.122.

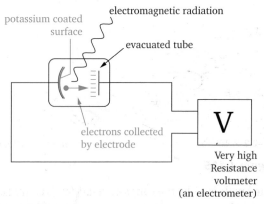

**Figure 4.122**

The photocell has two electrodes in an evacuated tube. One electrode is coated with potassium. Monochromatic light is incident on the potassium surface. Electrons are emitted. The electrons are collected by the collecting electrode, which becomes more negative as a result.

Eventually no more electrons can reach the collector. The potential difference $V$ between the two electrodes is measured using the high resistance voltmeter. The maximum kinetic energy of the electrons emitted by the radiation is $V$ electron volts or $eV$ joule.

The maximum kinetic energy can be measured for different frequencies of incident radiation (different photon energies). A graph of $E_{k(max)}$ against $hf$ should be similar to that in Figure 4.121, confirming the photoelectric equation.

EXAMPLE:

(a) For a magnesium surface, determine:

   (i)   the minimum photon energy required to remove an electron

   (ii)  the frequency of the photon.

(b) Calculate the maximum kinetic energy of an electron emitted when a tungsten surface is illuminated by light of wavelength 320 nm.

(a) (i)   When an electron is just emitted,
$$hf = \Phi$$
$$= 5.8 \times 10^{-19}\,\text{J}$$

   (ii)  When the electron is just emitted,
$$f = \frac{\Phi}{h} = \frac{5.8 \times 10^{-19}}{6.6 \times 10^{-34}} = 8.8 \times 10^{14}\,\text{Hz}$$

(b)

$$\text{photon energy} = \frac{hc}{\lambda}$$

$$= \frac{6.6 \times 10^{-34} \times 3.0 \times 10^{8}}{320 \times 10^{-9}}$$

$$= 6.2 \times 10^{-19}\,\text{J}$$

KE of electron = photon energy − work function

$$= (6.2 - 5.8) \times 10^{-19}\,\text{J}$$

$$= 0.4 \times 10^{-19}\,\text{J}$$

## Test your understanding

1 Complete the following table.
(Speed of electromagnetic radiation
$c = 3.0 \times 10^{8}\,\text{m s}^{-1}$.)

**Table 4.9**

|  | X-ray | Ultra-violet | Micro-wave | Radio-wave |
|---|---|---|---|---|
| $\lambda/\text{m}$ | $1.5 \times 10^{-10}$ |  |  | 200 |
| $f/\text{Hz}$ |  | $3.2 \times 10^{15}$ |  |  |
| photon energy/J |  |  | $5.0 \times 10^{-24}$ |  |

2 When light passes from one medium into another the frequency is unchanged. State and explain:
  (a) what happens to the wavelength
  (b) what happens to the photon energy
3 (a) State two reasons why electromagnetic radiation may be thought to be a wave.
  (b) State one reason why it may be thought to be a particle.
4 (a) Determine the maximum wavelength of the radiation that will produce the photoelectric effect using a potassium surface of work function $3.5 \times 10^{-19}$ J.
  (b) What frequency radiation will cause the emission of electrons of energy $2.0 \times 10^{-19}$ J using a potassium surface?
5 Draw a graph showing how the maximum kinetic energy of the electrons emitted by a zinc surface varies with photon energy.
6 The threshold frequency of a surface is $5.5 \times 10^{14}$ Hz.
  (a) Explain what is meant by *the threshold frequency*.
  (b) Calculate the wavelength of this radiation.
  (c) Calculate the photon energy of this radiation.

# Energy Levels

## Line spectrum of hydrogen

Figure 4.123 shows a schematic diagram of an arrangement for observing the visible spectrum of atomic hydrogen.

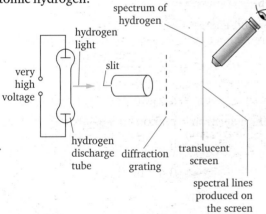

**Figure 4.123**

The spectrum consists of a series of coloured lines. Each colour in a spectrum is produced by a particular wavelength (or frequency).

The line spectrum is shown in Figure 4.124.

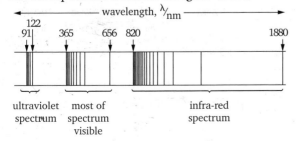

**Figure 4.124**

This shows that only certain well-defined frequencies of light are emitted by hydrogen. Only certain photon energies, those corresponding to the frequencies, are present in hydrogen light.

Atoms of a given element produce a line spectrum that is unique to that element. This can be used to analyse substances to determine which elements are present in it.

## Energy levels

The line spectrum can only be explained by a theory that suggests that the electrons in atoms can only exist in well-defined **energy levels**.

Since different atoms produce different line spectra, it follows that each atom has a unique set of energy levels.

Figure 4.125 shows the energy level diagram for atomic hydrogen. The energies are given in both J and eV.

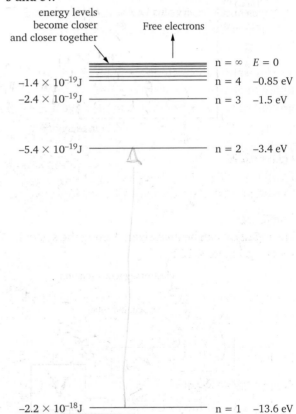

**Figure 4.125**

The levels are like the rungs on a ladder. You are able to stand on any rung but you cannot stand between the rungs. Similarly the electrons can be in any one of the levels, but cannot be anywhere in between.

Unlike the rungs of a ladder, however, note that the levels are not evenly spaced. The higher energy levels are closer together.

The energy levels are identified by numbers $n = 1$, 2, 3, ........ $\infty$, the lowest level having $n = 1$.

$n = 1$ is called the **ground state**. This is the lowest level that an electron can occupy in an atom.

The other levels are called **excited states**.

### Energy levels are negative

An electron that is **just free** of the atom has **zero energy**. Since energy has to be supplied to an electron in an energy level for it to become free, the energies of electrons in the atom are negative. The lower the energy level the more negative it is.

Energy has to be given to an electron in an atom to move it to a higher level or to liberate it from the atom.

## Ionisation

Ionisation is the process of removing an electron from an atom leaving the atom with a positive charge.

An electron in the ground state ($n = 1$) is the most tightly bound electron. An electron in the ground state needs the most energy to liberate it (i.e to ionise the atom).

The ionisation energy is different for atoms of different elements because the energy of the ground state varies from element to element.

## Excitation

Excitation is the process in which electrons are moved from a lower to a higher energy level.

The process in which an electron moves from one level to another is called a **transition**.

An electron in the ground state may be excited into any of the higher levels. An electron has to gain energy to do this. The energy needed is the difference in energy between the two levels.

In Figure 4.125 an electron in the ground state may be excited from the $n = 1$ into the $n = 2$ level.

The energy difference between these two levels is:
*lower energy = $-G$ always.*
$$-3.4 - (-13.6) = 10.2 \text{ eV or } 1.6 \times 10^{-18} \text{ J}$$

This energy has to be supplied for the transition to take place. The energy can be provided by:
- electrons with energy equal to or greater than the difference between the energy levels
- photons that have an energy equal to the difference between the two levels.

### Instability of excited states

An electron in an excited state can only stay there for a very short time. The excited state is **unstable**. Almost instantaneously, the electron moves into one of the lower energy states. This process is called **relaxation**.

The electron may move to the ground state in one step or it may do so in stages moving to one or more of the lower levels on the way. This is illustrated in Figure 4.126.

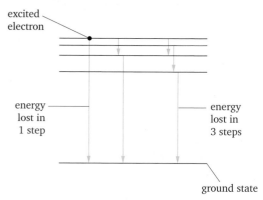

**Figure 4.126**

An electron in level $n = 4$ may move to the ground state by any one of the indicated routes.

For each transition to a lower energy level, the electron has to lose energy. This energy is in the form of a photon of electromagnetic radiation.

For a transition from a level with energy $E_2$ to one with a lower energy $E_1$, the energy of the emitted photon $hf$ is given by:

$$hf = E_2 - E_1$$

Since the energy levels are well defined for a given atom so are:
- the possible transitions
- the photon energies
- the frequencies of the electromagnetic radiation.

Hence, the existence of well-defined frequencies in the line spectrum of a given element provides evidence for the existence of energy levels in atoms. If the electron could have any energy, then all transitions would be possible and the spectrum would be a continuous spectrum rather than a line spectrum.

## Visible, Ultraviolet and Infrared Spectra

The energy of the photons and hence the type of electromagnetic radiation emitted depends on the energy difference between the levels involved in the transition.

Ultraviolet radiation requires a larger energy difference than visible light and infrared a lower energy difference.

Figure 4.127 shows transitions that take place into the $n = 1$, $n = 2$ and $n = 3$ levels for hydrogen.

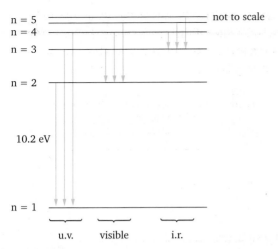

**Figure 4.127**

For transitions into the $n = 1$ level (the ground state) the smallest photon energy emitted is $1.6 \times 10^{-18}$ J which, using $\dfrac{hc}{\lambda}$, corresponds to a wavelength of $1.2 \times 10^{-7}$ m. The greatest photon energy is $2.2 \times 10^{-19}$ J, corresponding to a wavelength of $1.4 \times 10^{-7}$ m.

These wavelengths are in the ultraviolet part of the electromagnetic spectrum so all transitions to the $n = 1$ level produce radiation in the ultraviolet part of the spectrum.

For transitions into the $n = 2$ level, the range of possible wavelengths is $3.6 \times 10^{-7}$ m to $6.5 \times 10^{-7}$ m, most of which is in the visible part of the spectrum.

For transitions into the $n = 3$ level, the range is $8.0 \times 10^{-7}$ m to $1.9 \times 10^{-6}$ m which are all in the infrared part of the spectrum.

Electron transitions for heavy atoms such as copper and tungsten can be very large, giving rise to high energy photons in the X-ray part of the spectrum.

The mapping of energy levels for any element is a matter of determining the set of energy levels that can give rise to the particular wavelengths that are produced by the element.

## Absorption spectra

This is a spectrum produced when white light is passed through a vapour or gas. Some wavelengths are removed from the light so that the continuous spectrum produced is crossed by dark lines.

The missing wavelengths are due to the photons at these wavelengths having the right energy to produce excitation of the vapour or gas. Following excitation, the energy is re-emitted in all directions and at different wavelengths due to the various routes for relaxation to occur. The intensity at those wavelengths is therefore reduced.

## Test your understanding

1 Determine the difference between the energy levels that produces a photon energy of 500 nm.

2 State what is meant by:
   (a) the ground state of an electron in an atom
   (b) an excited state of an electron in an atom.

3 Explain the difference between *excitation* and *ionisation*.

4 A photon is emitted by a transition between two energy levels at $-1.5$ eV and $-3.4$ eV.
   $1$ eV $= 1.6 \times 10^{-19}$ J,
   $h = 6.6 \times 10^{-34}$ Js
   $c = 3.0 \times 10^8$ m s$^{-1}$
   (a) Explain why the energy levels are negative.
   (b) Calculate the frequency of the radiation emitted when an electron moves from the higher level into the lower level.
   (c) Calculate the wavelength of this radiation.

5 Figure 4.128 shows the energy level diagram for hydrogen.

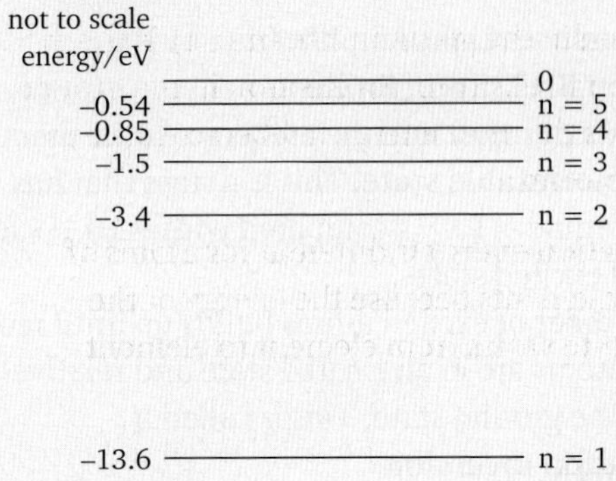

not to scale
energy/eV

| | |
|---|---|
| $-0.54$ | 0 |
| $-0.85$ | $n = 5$ |
| $-1.5$ | $n = 4$ |
| | $n = 3$ |
| $-3.4$ | $n = 2$ |
| | |
| $-13.6$ | $n = 1$ |

**Figure 4.128**

   (a) State the ionisation energy of hydrogen:
       (i) in eV
       (ii) in J.
   (b) Following excitation from level $n = 1$ to level $n = 5$, hydrogen relaxes by emitting photons.
       (i) Calculate the highest photon energy, in J, that may be emitted.
       (ii) Calculate the lowest photon energy, in J, that may be emitted.
       (iii) How many different photon energies are possible?

6 An X-ray of wavelength $1.1 \times 10^{-10}$ m is produced when an electron undergoes a transition between energy levels in a heavy atom. Calculate the difference in energies between these levels.

7 The spectrum of an atom contains the following three wavelengths:
   $$4.1 \times 10^{-7} \text{ m}$$
   $$5.1 \times 10^{-7} \text{ m}$$
   $$22 \times 10^{-7} \text{ m}.$$
   These wavelengths are known to come from transitions from two excited states into an energy level at $-8.2 \times 10^{-19}$ J.
   (a) Show on an energy level diagram how this is possible.
   (b) Determine the energies of the photons emitted.
   (c) Determine the energies of the other two levels.

# Lasers

### Operation of a normal discharge tube

When electricity is discharged through a gas in a discharge tube:

- Energy to excite the atoms is provided by collisions with electrons.
- At any time most atoms are in the ground state and relatively few are excited.
- Excited electrons fall spontaneously to the ground state emitting photons.
- Photons are emitted randomly and in all directions.
- Energy radiated in any particular direction is weak.
- Radiated waves have different phases so the light from the different atoms is incoherent.

## Operation of a Laser

A laser produces **L**ight **A**mplification by the **S**timulated **E**mission of **R**adiation.

- Energy is pumped into the atoms to excite them into a **metastable state**. This is a state that has a long lifetime $\approx 10^{-3}$ s compared with $\approx 10^{-8}$ s for normal excited states.
- The number of excited atoms builds up until most of the atoms are in an excited state and relatively few in the ground state. This is called a **population inversion**.
- An incident photon with the correct frequency **stimulates** the atoms so that the excited electrons all move to a lower level at the same instant.
- The emitted photons are all in the same direction and have the same phase as the incident photon and so are **coherent**.
- Because all the emissions occur at the same time and in the same direction there is a **high intensity beam in a particular direction**.

## Operation of a Helium–Neon Laser

Figure 4.129 shows a simplified diagram of a helium–neon laser.

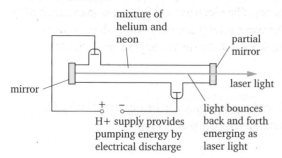

**Figure 4.129**

Figure 4.130 shows the energy levels that give rise to the laser action. (Note that, for simplicity, the energies are given relative to the ground state rather than being given their actual negative energies.)

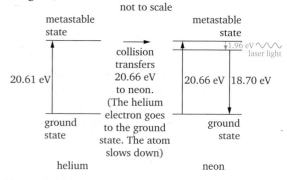

**Figure 4.130**

Helium atoms are pumped into a state that is 20.61 eV above the ground state. This state, called a **metastable state**, exists for a long time.

Neon atoms are now excited into the level that is 20.66 eV above their ground state. This state also exists for a long time.

The energy to excite the neon atoms comes from collisions with the excited helium atoms:

- 20.61 eV comes from the excited electron in the helium
- 0.05 eV of the energy comes from KE of the helium atoms.

In normal circumstances, there would be many more atoms in the lower energy state than in the higher energy state (see Figure 4.131). However, in the laser there is a **population inversion** such that there are more neon atoms in the 20.66 eV level than in the level that is 18.70 eV above the ground state.

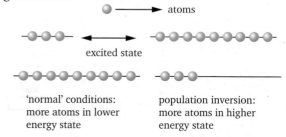

**Figure 4.131**

A photon emitted spontaneously by a neon atom (moving from the 20.66 eV to the 18.70 eV level) now stimulates all the excited neon atoms to do the same. There is, therefore, an intense beam of coherent photons of energy 1.96 eV. These photons are in the red part of the spectrum and have a wavelength of 632.8 nm.

## Uses of Lasers

In a physics laboratory, low power lasers make measurements in diffraction and interference experiments relatively easy. Low power laser pens are commonly used by lecturers to point to important features on an image on a screen.

### Safety note

The intense beam of even low power lasers can cause serious damage to the eyes. Take care never to look at a laser beam directly or to direct the beam at others. Reflections from highly reflecting surfaces are as dangerous as the original beam so beware of such reflections when setting up experiments.

### Medical uses

Because of their narrow intense beam, lasers are used to destroy tissue on a localised area. They can be used to drill a small 'pin-hole' to restore the sight of people inflicted with cataracts. They also can be used to weld a detached retina or break up gall stones or kidney stones.

### Industrial uses

Lasers are used to weld metals together or to cut through metal sheets. Because they produce a beam that is straight and can be seen over very large distances, they are useful in surveying.

### Application in communications

Modulated laser beams are used to convey information down optical fibres. Lasers are used to produce the pits, which represent bits in digital recording using a compact disc (CD), and to read the information from the CD.

A diagram showing the arrangement for reading from a CD is shown in Figure 4.132.

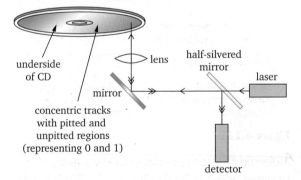

**Figure 4.132**

When a CD is recording information, pits are formed to represent a binary 1. Where there is no pitting there is a binary 0. A narrow beam of light from a solid-state laser is directed onto the CD. When it lands on a pit, little light is reflected and the detector recognises the bit as a 1. Where there is no pit the intensity of light reflected is high and the detector recognises a 0.

### Test your understanding

1  A ruby laser consists of a ruby rod crystal that is aluminium oxide. Some of the aluminium atoms in the ruby are replaced by chromium.

Figure 4.133 shows the chromium energy levels that produce the laser action.

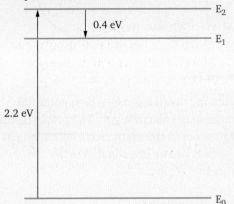

**Figure 4.133**

The laser light has a photon energy of 1.8 eV.
  (a)  Which of the levels is a metastable state?
  (b)  What is the photon energy that will stimulate the emission?
  (c)  What is the frequency of the light from a ruby laser?

2  State two differences between the light from a 2 mW filament light source and the light from a 2 mW laser.

3  The beam of light from a 10 mW helium–neon laser is 0.50 mm diameter.
  (a)  Calculate the intensity of the beam in mW m$^{-2}$.
  (b)  Calculate the number of photons emitted per second by the laser.
  (c)  For a distance of 0.1 m calculate the ratio:

$$\frac{\text{intensity of light from the laser}}{\text{intensity of light from a 10 mW point light source}}$$

## Wave–Particle Duality of Matter

All early experiments to establish the properties of electrons showed that electrons are particles with mass and charge. However, observations in later experiments can only be explained using a wave theory. The electrons exhibit wave–particle duality. The wave effects are only noticeable for small particles but the theory applies to all matter.

### Electron properties

Electron properties can be observed using the arrangement shown in Figure 4.134.

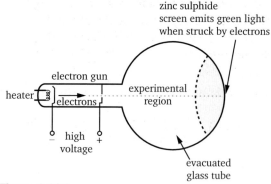

**Figure 4.134**

Electrons are accelerated by a high potential difference in the electron gun. The properties of the electron beam (a stream of electrons moving at the same speed) can be investigated in the experimental region.

### Particle behaviour

Electrons appear to travel in a straight line when they meet an obstacle casting a sharp shadow such as that in the Maltese cross experiment (Figure 4.135). No diffraction is observed as would be expected with waves.

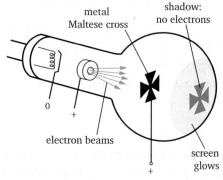

**Figure 4.135**

All electrons of a given speed are deflected in the same way by a magnet or by an electric field as shown in Figure 4.136.

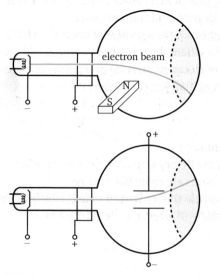

**Figure 4.136**

Electrons can be accelerated to higher speeds by increasing the force acting on them, such as by using a higher accelerating potential difference in an electron gun. The electrons then have higher energy and momentum.

Electrons can lose energy gradually in collisions with gas atoms (producing ionisation or excitation) so that their speed decreases. This causes the spiral path seen in cloud chambers (see Figure 4.137).

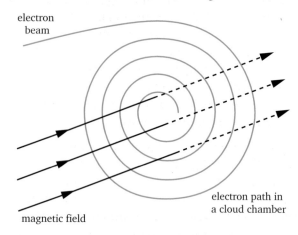

**Figure 4.137**

Experiments studied in detail in Module 5 show that the electrons behave as particles of mass $9.1 \times 10^{-31}$ kg, each electron carrying a charge of $1.6 \times 10^{-19}$ C. A wave would not carry a charge.

## Wave Behaviour

This can be demonstrated using an evacuated tube in which a beam of electrons is incident on a thin sheet of graphite as shown in Figure 4.138.

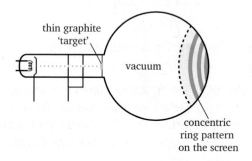

**Figure 4.138**

A particle theory predicts that the screen should show a single blurred spot where the electrons hit the screen. The blurring would be due to collisions of the electrons with matter in the graphite.

In fact, the pattern on the screen is as shown in Figure 4.139.

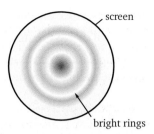

screen

bright rings

**Figure 4.139**

The pattern of circles showing bright and dark rings is observed. This is similar to the interference pattern that is observed when a similar experiment is performed using X-rays. The interference pattern is explained by applying the wave theory. The regular atomic layers in the graphite diffract the waves, which interfere to produce positions of maximum and minimum intensity.

When electrons strike the phosphor in the diffraction tube, light is emitted. The pattern of circles shows that electrons passing through the graphite can only travel to certain parts of the screen. As with X-rays, such a pattern can only be explained using a wave theory.

If the electrons are accelerated to a higher speed, the circles of maximum brightness have a smaller radius. From the work on diffraction studied in the AS course, you should appreciate that the electron waves have a shorter wavelength when the speed is increased.

### De Broglie equation

De Broglie proposed that the wavelength associated with a particle is given by:

$$\lambda = \frac{h}{mv}$$

where    $h$ is Planck's constant ($6.6 \times 10^{-34}$ J s)
         $mv$ is the momentum of the electron ($p$).

Particles are made up of vast numbers of atoms and hence have relatively large masses. Even slow moving particles therefore have a very large momentum and hence a very small wavelength.

For example, a ball of mass 0.25 kg moving at a speed of only 5.0 m s$^{-1}$ has a momentum of 1.25 kg m s$^{-1}$, so the wavelength is $5.3 \times 10^{-34}$ m. Even a small dust particle of mass $1 \times 10^{-9}$ kg ($1\mu g$) moving at 1 m s$^{-1}$ has a wavelength of $6.6 \times 10^{-25}$ m.

No experiment can be devised to show diffraction effects using such a small wavelength. However, when the particles have very low mass it is possible to produce longer wavelengths and diffraction effects are observable. A wavelength of about $1 \times 10^{-10}$ m, for example, can produce the diffraction effects observable in the electron diffraction tube.

**EXAMPLE:** An electron of mass $9.1 \times 10^{-31}$ kg is accelerated to a speed of $3.0 \times 10^7$ m s$^{-1}$. Calculate the wavelength associated with the electron.

$p = mv$
$\quad = 9.1 \times 10^{-31} \times 3.0 \times 10^7 = 2.7 \times 10^{-25}$ kg m s$^{-1}$

wavelength $\lambda = \dfrac{h}{p} = \dfrac{6.6 \times 10^{-34}}{2.7 \times 10^{-25}} = 2.4 \times 10^{-10}$ m

### Demonstrating that $\lambda \infty \frac{1}{p}$

This relationship can be demonstrated using an electron diffraction tube.

The electron momentum, $p = \sqrt{(2mE_k)}$, where $E_k$ is the kinetic energy of the electron.

In an electron gun, the kinetic energy is proportional to the accelerating potential difference. Hence:    $E_k \propto V$

$$\lambda \propto \frac{1}{\sqrt{V}}$$

The wavelength is proportional to the spacing of adjacent maxima (as when using a diffraction grating with visible light).

The experiment shows that when the accelerating potential difference is doubled, the diameter $d$ of the first ring (the first order maximum) decreases by $1/\sqrt{2}$, i.e. to 0.71 of the original radius, as expected.

### Meaning of the amplitude

The probability of finding a particle at a point is proportional to the square of the amplitude of the wave at that point:

$$\text{Probability} \propto \text{amplitude}^2$$

A stream of electrons, all travelling at the same speed, produces a higher intensity where the wave has a larger amplitude because:
- electrons have a greater chance of arriving at a point where there is a larger amplitude
- more electrons arriving at a point means more energy arriving per second.

## Summary

In summary:
- a wave theory is used to determine where electrons are more likely to go
- a particle theory is used to describe how energy is exchanged when the electron arrives.

## Energy Levels and Wave–Particle Duality

Application of wave–particle duality to the electrons enabled Schrödinger to produce a wave equation that predicts the electron energy levels in a hydrogen atom. What follows is a simplified model to show how wave-particle duality can by applied. An electron in an atom must exist between the nucleus and the 'edge of the atom'. This means that:

- the probability of finding an electron at the centre and at the edge must be zero
- the electron wave is trapped between the two boundaries just as the wave on a stringed instrument is trapped between the fixed ends
- an electron standing wave is set up between the two limits as shown in Figure 4.140
- only waves of well defined wavelength in which a whole number of half wavelengths fits between the two limits are possible, as shown in Figure 4.140
- since a defined wavelength is a defined energy the electron can only take certain well defined (or quantised) energies in the atom
- transitions of electrons between high and low energy levels give rise to the observed line spectrum of hydrogen.

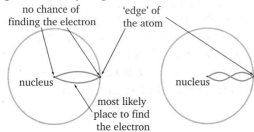

**Figure 4.140**

The most probable place to find an electron is at the antinode of the standing wave, where the amplitude is greatest.

## Size of an Atom

Assume the radius of a hydrogen atom to be about $1 \times 10^{-10}$ m.

The simplest standing wave (the first harmonic) produced by the electron wave has a wavelength of $2 \times 10^{-10}$ m.

Using
$$\lambda = \frac{h}{\sqrt{2mE_k}}$$

gives a value for $E_k$ of $6 \times 10^{-18}$ J.

At the radius assumed, the energy needed for an electron to escape from the nucleus is $2.3 \times 10^{-18}$ J, so an electron with the wavelength assumed would not be held in the atom.

However, the order of magnitude is the same and suggests that the theory is on the right track.

If the electron is assumed to be in a larger radius, the wavelength is increased so the kinetic energy decreases. In this model, the radius would be that in which the kinetic energy of the electron matches exactly the energy required to remove the electron from the atom.

The energy needed to remove an electron from the nucleus is proportional to $\frac{1}{r}$. You may like to try to show that this simple model gives a radius $2.6 \times 10^{-10}$ m for the radius of the atom.

**Test your understanding**

1  An electron has a mass of $9.1 \times 10^{-31}$ kg. The Planck constant is $6.6 \times 10^{-34}$ J s. Calculate:
   (a) the de Broglie wavelength of an electron that has a speed of $1.6 \times 10^6$ m s$^{-1}$
   (b) the momentum of an electron that has a kinetic energy of $4.1 \times 10^{-16}$ J
   (c) the kinetic energy of an electron of wavelength $1.5 \times 10^{-10}$ m.

2  Calculate the speed of a person of mass 70 kg who has the same wavelength as an electron with a speed of $3.0 \times 10^5$ m s$^{-1}$.

3  A neutron of mass $1.7 \times 10^{-27}$ kg has a wavelength of $5.0 \times 10^{-13}$ m. Calculate:
   (a) its momentum
   (b) its speed
   (c) its kinetic energy.

4  Calculate the minimum kinetic energy that an electron would have if it were to form a standing wave in a nucleus of diameter $1.5 \times 10^{-13}$ m.

5  Figure 4.141 shows the stationary wave of an electron trapped in an atom, using a simplified model.

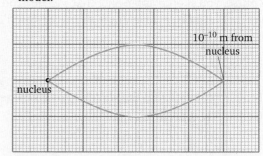

**Figure 4.141**

   (a) Explain where the electron is most likely to be found in this model.
   (b) Where will the probability of finding the electron be half that at the most likely position?

6  For an alpha particle to be captured by a nucleus its wavelength must be less than the nuclear diameter ($1.5 \times 10^{-14}$ m). An alpha particle has a mass of $6.8 \times 10^{-27}$ kg.
   Calculate the minimum kinetic energy of an alpha particle that can be captured:
   (a) in J
   (b) in MeV.

7  (Synoptic question; refer to AS Module 2)
   Electrons travelling at a speed of $3 \times 10^6$ m s$^{-1}$ are incident on a diffraction grating with 150 lines per mm.
   (Mass of an electron $= 9.1 \times 10^{-31}$ kg, Planck constant $h = 6.6 \times 10^{-34}$ J s.)
   (a) Calculate the electron wavelength.
   (b) Explain whether the fringes produced by the diffraction grating would be visible with the naked eye.

# Unit 5 Electric and Gravitational Fields

## Electric Fields

An electric field is a region in which a charged body will experience a force. Electric fields are also produced by charged bodies: the dome of a Van der Graaff generator creates an electric field around it. A pair of parallel plates with a potential difference between them will also create an electric field.

### Using electric field lines to draw electric fields

A field line shows the direction of the force that would be experienced by a small positive charge if it were placed at that point in an electric field. Figure 5.1 shows the electric field near a positively charged sphere.

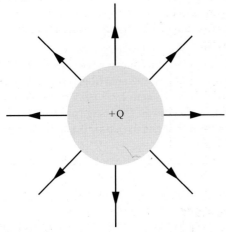

**Figure 5.1**

Notice:
- the field lines all have arrows. Electric field strength is a vector quantity and thus has magnitude and direction
- All the field lines are shown with arrows pointing away from the sphere as a small positive test charge would be repelled from the sphere
- the lines are closer together near to the sphere. This indicates that the field is stronger near to the sphere.

Figure 5.2 shows the electric field between two parallel metal plates with a potential difference between them.

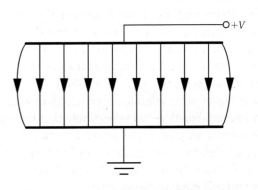

**Figure 5.2**

Notice:
- the field lines are evenly spaced and parallel, showing that the electric field is uniform. This means that the field is equally strong and has the same direction at all points.
- at the very edges of the field, it stops being uniform. This is called an edge effect.

## Uniform Electric Fields

The main measure of the "effectiveness" of an electric field is the **electric field strength**, $E$. Electric field strength at a point in a field is defined as the force that would be experienced, per unit charge, by a small positive test charge placed at that point in the field. Thus, if a test charge of $Q$ coulomb is placed in the field and experiences a force $F$, the field strength is:

$$E = \frac{F}{Q}$$

alternatively: $\qquad F = EQ$

### Work done moving charges in uniform electric fields

An electric field is created between two parallel conducting plates separated by a distance $d$ and having a potential difference of $V$ between them. The situation is illustrated in Figure 5.3, together with a graph showing the variation of potential with position between the plates.

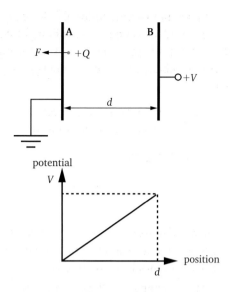

**Figure 5.3**

The charge $Q$ experiences a force $F$ due to the electric field. To move $Q$ from **A** to **B**, a force of the same magnitude would be needed. The work done moving $Q$ from **A** to **B** is $W$.

$$W = Fd$$

but: $$F = EQ$$

so, $$W = EQd$$

but potential difference, $V$, is defined as work done per unit charge:

$$V = \frac{W}{Q}$$

so: $$V = \frac{W}{Q} = \frac{EQd}{Q}$$

$$V = Ed$$

or $$E = \frac{V}{d}$$

This equation applies to uniform electric fields only. It is not a definition of electric field strength.

It is hardly surprising that the electric field strength increases if the potential difference increases or if the distance between the plates decreases.

Notice also that the field strength is numerically equal to the gradient of the graph of potential against position. This fact is correct for all electric fields including non-uniform fields (see Figure 5.19 on page 93).

(see Figure 5.19 on page 93).

**Test your understanding**

1  The table below applies to uniform electric fields. Fill in the gaps.

**Table 5.1**

|   | $E$ | $V$ | $d$ |
|---|---|---|---|
| a |  | 400 V | 0.080 m |
| b | $6.0 \times 10^5 \, \text{V m}^{-1}$ | 20 kV |  |
| c | $8.2 \times 10^4 \, \text{V m}^{-1}$ |  | 12 cm |

2  The table below applies to point charges in uniform electric fields. Fill in the gaps.

**Table 5.2**

|   | $F/\text{N}$ | $E$ | $Q/\text{C}$ |
|---|---|---|---|
| a |  | $6300 \, \text{V m}^{-1}$ | $1.6 \times 10^{-19}$ |
| b | $3.4 \times 10^{-13}$ | $5000 \, \text{V m}^{-1}$ |  |
| c | $8.7 \times 10^{-5}$ |  | $6.4 \times 10^{-9}$ |

3  Two plates have a potential difference of 4500 V across them and are separated by a distance of 0.064 m. A charged body of $+3.2 \times 10^{-17}$ C is moved from the negative plate to the positive plate.
Calculate:
(a) the electric field strength of the field
(b) the force on the charged body
(c) the work done moving the charged body.

4  An electron is accelerated from rest by an electric field of field strength $1.3 \times 10^4 \, \text{V m}^{-1}$ over a distance of 0.84 m. The charge on the electrons is $1.6 \times 10^{-19}$ C and its mass is $9.1 \times 10^{-31}$ kg.
Calculate:
(a) the potential difference over which the electron is accelerated.
(b) the final kinetic energy of the electron
(c) the final velocity of the electron.

5  (a) Which of these electric field parameters are vectors: $F$, $E_p$, $E$, $V$.
(b) Give units for the parameters mentioned in (a).

# Gravitational Fields

A gravitational field is a region in which an object that has mass experiences a force. Gravitational fields are created by any object that has mass. It is not usual to draw gravitational fields using gravitational field lines; however, by comparison with electric field lines, the Earth's gravitational field would appear to be like that drawn in Figure 5.4.

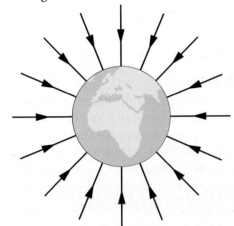

**Figure 5.4**

Notice that the arrows all point towards the Earth: gravitational forces are always attractive – there is no gravitational repulsion.

There is no way of producing a uniform gravitational field similar to the way in which parallel metal plates produce a uniform electric field. However, when viewed on a small scale, the gravitational field near the Earth's surface is approximately uniform. See Figure 5.5.

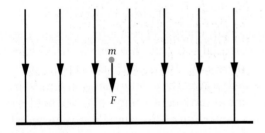

**Figure 5.5**

Gravitational field strength is given the symbol $g$ and is defined as the force acting per unit mass on a body. So, for a body of mass m:

$$g = \frac{F}{m}$$

or

$$F = mg$$

A body of mass $m$ in a gravitational field of field strength $g$ would also need a force of magnitude $mg$ in order to lift it. When the body is lifted through a height $\Delta h$, the change in potential energy $\Delta E_p$ is given by:

$$\Delta E_p = \text{work done lifting body}$$
$$= \text{force} \times \text{distance moved}$$

$$\Delta E_p = mg\,\Delta h$$

## Potential in uniform gravitational and electric fields

When a body of mass $m$ is lifted by a distance $h$ in a gravitational field of field strength $g$, the change in gravitational potential $\Delta V_G$ is given by:

$$\Delta V_G = \text{change in PE per unit mass}$$

$$= \frac{mg\,\Delta h}{m}$$

$$\Delta V_G = g\,\Delta h$$

Notice that, as the field is uniform, the change in gravitational potential is simply proportional to the height through which the body is raised. The change in potential with distance can be illustrated by using lines of equipotential. This is a line joining points of equal potential. They are similar to the contours on a map that join points of equal height (see Figure 5.6).

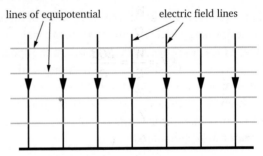

**Figure 5.6**

Notice that the lines of equipotential are evenly spaced in the uniform gravitational field. Notice also that they are at right angles to the gravitational field lines. The lines are only evenly spaced when the field is uniform. Similarly, the equation $\Delta E_p = mg\,\Delta h$ is only valid when the field can be considered to be uniform. As far as the Earth's magnetic field is concerned, this holds true when $\Delta h$ is small compared with the radius of the Earth.

Lines of equipotential can also be drawn to illustrate electric fields (see Figure 5.7).

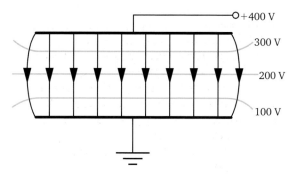

**Figure 5.7**

Notice that, once again, the lines of equipotential are evenly spaced where the field is uniform. Notice too that the lines of equipotential are perpendicular to the electric field lines, even where the field is non-uniform at the edges.

## Examples of energy conservation in gravitational fields

When a body of mass $m$ falls from rest through a gravitational field, the change in potential energy ($\Delta E_p$) is equal to the change in kinetic energy ($\Delta E_k$), providing that the body does not have to do any work against resistance forces such as viscous drag (see Module 1).

$$\Delta E_p = \Delta E_k$$

$$mg\,\Delta h = \tfrac{1}{2}mv^2$$

When a falling body hits the ground, some or all of its kinetic energy will be converted to internal energy of the body and of the ground. In the special case of water falling down a waterfall into a still pool, it may be assumed that all of the water's kinetic energy changes into internal energy of the water. This enables the temperature rise, $\Delta\theta$, of the water to be calculated.

$$\Delta E_k = \Delta(\text{internal energy})$$

$$mg\,\Delta h = mc\,\Delta\theta$$

where c = specific heat capacity of the water

$$\Delta\theta = \frac{g\,\Delta h}{c}$$

## Conservative fields

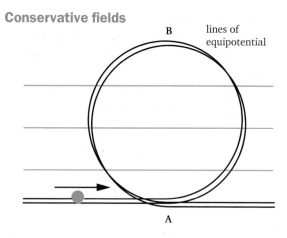

**Figure 5.8**

Figure 5.8 shows a marble running along a frictionless track. Between **A** and **B**, some of its kinetic energy is converted into potential energy. As it moves back to **A**, it loses an equal amount of potential energy. Since there has been no net change in potential energy, the final kinetic energy is equal to the initial kinetic energy.

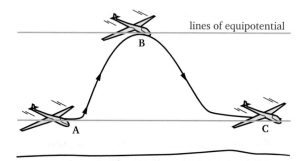

**Figure 5.9**

In Figure 5.9, the glider climbs from **A** to **B**, losing kinetic energy as it gains potential energy. As it moves from **B** to **C** it loses potential energy and gains kinetic energy. Since **A** is on the same line of equipotential as **C**, there has been no net change in potential energy. Therefore, in the absence of resistance forces, the kinetic energy at **C** must be the same as the kinetic energy at **A**. The field is said to be a *conservative* field, since there is no overall change in energy when a body moves from point **A** to point **C** if both are at the same potential.

The same concept applies equally to electric fields. In Figure 5.10, no net energy is required to move Q from **A** to **B** by any route since both points are at the same potential.

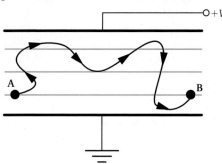

**Figure 5.10**

## Units for field strength

For electric fields:

$$E = \frac{F}{Q} \text{ so } E \text{ is measured in N C}^{-1}$$

For gravitational fields:

$$g = \frac{F}{m} \text{ so } g \text{ is measured in N kg}^{-1}$$

### Test your understanding

1 The table relates to forces on masses in uniform gravitational fields. Fill in the gaps:

**Table 5.3**

|   | **F** | **m** | **g** |
|---|---|---|---|
| a | 170 N | 17.3 kg | |
| b | | 17.3 kg | 1.6 N kg$^{-1}$ |
| c | 0.043 N | | 82 N kg$^{-1}$ |

2 A body of mass 16 kg is lifted from **A** to **B**, a distance of 12 m in a field of field strength 9.8 N kg$^{-1}$. Calculate:
   (a) the work done lifting the body
   (b) the change in potential energy of the body
   (c) the difference in gravitational potential between points **A** and **B**.
3 A body of mass 86 kg falls 200 m. The gravitational field strength is 9.8 N kg$^{-1}$. Calculate:
   (a) the body's change in potential energy
   (b) the body's speed just before it hits the ground.
4 Rubber falls off a conveyor belt at a rate of 68 kg s$^{-1}$. It falls 3.7 m to the ground. The gravitational field strength is 9.8 N kg$^{-1}$.

Calculate:
   (a) the rate of change of potential energy
   (b) the change in temperature of the rubber if 15% of the change in potential energy goes towards raising its temperature. The specific heat capacity of rubber is 2100 J kg$^{-1}$ K$^{-1}$.
5 (a) Name the following parameters and state which of them are vectors: $F$, $E_p$, $g$, $V_G$.
   (b) Give units for the parameters mentioned in (a).

## Non-Uniform or Radial Fields

Whether the radial fields being considered are electric fields or gravitational fields, the important parameters are the same as for the appropriate uniform field as shown in Table 5.4.

**Table 5.4**

| Parameter | Electric field | Gravitational field |
|---|---|---|
| Force | $F$ | $F$ |
| Field strength | $E$ | $g$ |
| Potential energy | $E_p$ | $E_p$ |
| Potential | $V_E$ | $V_G$ |

### Radial electric fields

These are produced by point charges or by charges distributed around conducting spheres (see figures 5.11 and 5.12).

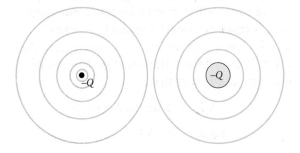

**Figure 5.11**

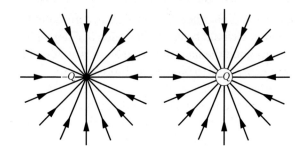

**Figure 5.12**

Note:
- the fields are identical for similar charges for radii greater than the radius, $r$, of the charged sphere.
- the field produced by the point charge has a greater field strength than the field produced by the sphere, for the values of radius less than $r$. Stronger fields have less distance between lines of equipotential or between the electric field lines.
- the fields get weaker as $r$ gets bigger.

## Force and field strength in radial electric fields

Coulomb was the first to investigate the factors affecting force in radial electric fields. He found that the force between two charged spheres (see Figure 5.13) was:
- directly proportional to the product of the charges
- inversely proportional to the square of the separation of their centres.

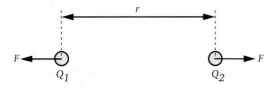

**Figure 5.13**

These relationships can be combined to give:

$$F = \frac{kQ_1Q_2}{r^2}$$

where $k$ is a constant.

If the **SI** system of units is being used:

$$k = \frac{1}{4\pi\varepsilon_o}$$

$\varepsilon_o$ is a constant known as the permittivity of free space (see also page 67). The value of $\varepsilon_o$ is $8.9 \times 10^{-12}$ F m$^{-1}$. If the field is being propagated in any material other than a vacuum, a different constant $\varepsilon$ would be used. The value of $\varepsilon$ depends on how good the material or medium is at allowing the electric field to propagate through it. All materials have values of $\varepsilon$ greater than $\varepsilon_o$. The value of $\varepsilon$ for air is, however, not much different from the value for free space (vacuum), so the two are usually taken to be the same. The equations above can be combined to give:

$$F = \frac{Q_1Q_2}{4\pi\varepsilon_o r^2}$$

Consider a field created by a point charge $Q$. A small charge $q$ is placed at point **A** in the field created by $Q$, a distance $r$ from $Q$ (see Figure 5.14).

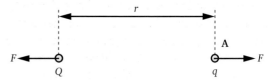

**Figure 5.14**

The force on $q$ caused by the field is given by:

$$F = \frac{Qq}{4\pi\varepsilon_o r^2}$$

The field strength at A caused by $Q$ is given by:

$$E = \text{force per unit charge} = \frac{F}{q}$$

$$E = \frac{F}{q} = \frac{Q}{4\pi\varepsilon_o r^2}$$

Figure 5.15 shows the variation of field strength with distance for the field around a point charge.

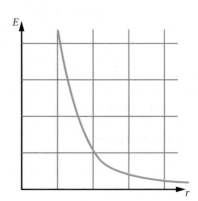

**Figure 5.15**

Notice that the graph shows an inverse square law relationship between field strength and distance.

Figure 5.16 shows the variation of field strength with distance from the centre of a hollow sphere carrying the same charge $Q$.

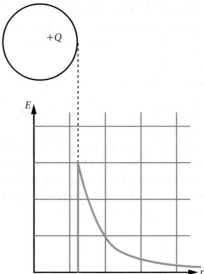

**Figure 5.16**

Notice that the field strength within the hollow sphere is zero, but that the graph is otherwise identical to Figure 5.15. Field strength is always zero within a hollow, conducting, charged object, e.g. within the dome of a Van der Graaff generator. This is why a car is a relatively safe place to be in a thunderstorm. Even if hit by lightning, no electric field exists within the hollow shell of the car.

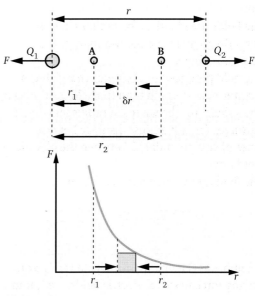

**Figure 5.17**

- When $r$ is infinite, the force between the two charged bodies is zero and the potential energy stored in the field is zero.
- If $Q_2$ is moved from infinity, at first only small amounts of work need to be done. As $r$ gets smaller, greater amounts of work need to be done.
- When work is done, the potential energy stored in the field is increased by an equal amount (releasing $Q_2$ would allow it to accelerate away, converting its potential energy to kinetic energy).
- As part of its movement from **A** to **B**, $Q_2$ moves through the small distance $\delta r$. This requires a small amount of work $\delta W$ to be done.

$$\delta W = \text{force} \times \text{distance moved}$$

magnitude of $\delta W = F\,\delta r$

$F\,\delta r$ is also the area of the rectangular column shown on the graph (ignoring the tiny triangle at the top).

- To move $Q_2$ all the way from **A** to **B** it would take many such moves, each needing small inputs of work $\delta W$. Each $\delta W$ is equivalent to the area under the appropriate part of the graph. So the whole work done moving $Q_2$ from **A** to **B** is equivalent to the whole area under the graph for values of $r$ between $r_1$ and $r_2$.
- The change in potential energy stored in the field when $Q_2$ is moved from **A** to **B** is also equal to the same area.

In these questions, $\varepsilon_o = 8.9 \times 10^{-12}\,\mathrm{F\,m^{-1}}$

1  Calculate the force between a proton and an electron separated be a distance of $6.3 \times 10^{-11}$ m. (Charge on an electron $= -1.6 \times 10^{-19}$ C, charge on a proton $= +1.6 \times 10^{-19}$ C.)

2  An alpha particle (charge $3.2 \times 10^{-19}$ C) approaches to a distance of $1.5 \times 10^{-15}$ m from a gold nucleus (charge $1.3 \times 10^{-17}$ C). Calculate:
 (a) the strength of the field created by the gold nucleus at the distance of the alpha particle's closest approach;
 (b) the force of repulsion between the alpha particle and the gold nucleus;
 (c) the acceleration of the alpha particle. (Mass of alpha particle $= 3.3 \times 10^{-25}$ kg.)

3  The field strength at the surface of the dome of a Van der Graaff generator is $2.1 \times 10^5$ V m$^{-1}$. The radius of the dome is 0.15 m. Calculate:
 (a) the charge on the dome;
 (b) the force experienced by an electron at the surface of the dome;
 (c) The force experienced by a $1.3 \times 10^{-8}$ C charge 0.40 m from the surface of the dome.

4  Two charges are separated by a distance of $4.3 \times 10^{-9}$ m. The force between them is $2.8 \times 10^{-8}$ N.
 (a) What will be the force between them at separations of $8.6 \times 10^{-9}$ m and $1.29 \times 10^{-8}$ m?
 (b) The potential energy stored in the field between the charges is $1.2 \times 10^{-16}$ J when they are at a separation of $4.3 \times 10^{-9}$ m. What will be the energies stored in the field when they are at separations of $8.6 \times 10^{-9}$ m and $1.29 \times 10^{-8}$ m?

## Potential energy stored in a radial electric field

- In Figure 5.17, when $Q_2$ is moved towards $Q_1$, work has to be done to overcome the repulsion between the charges. The work done moving $Q_2$ from **A** to **B** is $W$. Figure 5.17 also shows the variation in $F$ with separation, $r$.

### Determining potential energy changes from the graph

Figure 5.18 shows the variation of the force between the nucleus and electron of a hydrogen atom. The change in potential energy can be found by counting the squares under the graph and by determining the potential energy equivalent to each square, e.g. the shaded square in Figure 5.18.

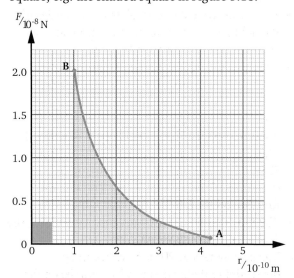

**Figure 5.18**

As the electron moves from $r = 4.2 \times 10^{-10}$ m to $r = 1.1 \times 10^{-10}$ m, its potential energy falls.

$\Delta E_\text{p}$ = area under the graph

= number of squares × 'value' of each square.

number of squares $= 14$

'value' of each square $= 0.5 \times 10^{-10}$ m
$\times 0.25 \times 10^{-8}$ N
$= 0.125 \times 10^{-18}$ Nm
$= 0.125 \times 10^{-18}$ J

$$\Delta E_\text{p} = 14 \times 0.125 \times 10^{-18}$$
$$= 1.8 \times 10^{-18} \text{ J}$$

As the electron makes this transition, it will have to lose this energy in some way and give out a photon of electromagnetic radiation (see Module 4). When considering the photon energy, it is more usual to quote the change in potential energy in eV.

$$\Delta E_\text{p} = \frac{1.8 \times 10^{-18}}{1.6 \times 10^{-19}}$$
$$= 11 \text{ eV}$$

### Calculating potential energy change

W is the sum of all the small inputs of work δW.

$$\delta W = F \delta r$$

so:

$$\delta W = \frac{Q_1 Q_2}{4 \varepsilon_o \pi r^2} . \delta r$$

so:

$$W = \sum \frac{Q_1 Q_2}{4 \varepsilon_o \pi r^2} . \delta r$$

The only way of performing this summation, apart from mechanically estimating the area under a graph, is to use integration. You are not required to be able to perform integration for this course, but you do need to know this particular result:

$$W = \frac{Q_1 Q_2}{4 \pi \varepsilon_o r}$$

As Potential Energy at a point is equivalent to the work done moving the charged body from infinity to that point:

$$E_\text{p} = W = \frac{Q_1 Q_2}{4 \pi \varepsilon_o r}$$

For the example of the movement of an electron within a hydrogen atom, it is the *change* in potential energy that is required as $Q_2$ moves from $r_1$ to $r_2$:

$$\Delta E_\text{p} = \frac{Q_1 Q_2}{4 \pi \varepsilon_o r_1} - \frac{Q_1 Q_2}{4 \pi \varepsilon_o r_2}$$

or

$$\Delta E_\text{p} = \frac{Q_1 Q_2}{4 \pi \varepsilon_o} \left( \frac{1}{r_1} - \frac{1}{r_2} \right)$$

Applying this equation to the above example:

$Q_1 = +1.6 \times 10^{-19}$ C (proton)
$Q_2 = -1.6 \times 10^{-19}$ C (electron)
$\varepsilon_o = 8.9 \times 10^{-12}$ Fm$^{-1}$
$r_1 = 4.24 \times 10^{-10}$ m
$r_2 = 1.06 \times 10^{-10}$ m

$$\Delta E_\text{p} = \frac{(1.6 \times 10^{-19})(-1.6 \times 10^{-19})}{4 \pi \times 8.9 \times 10^{-12}}$$
$$\times \left( \frac{1}{4.24 \times 10^{-10}} - \frac{1}{1.06 \times 10^{-10}} \right)$$

$$= \frac{-2.56 \times 10^{-38}}{1.12 \times 10^{-10}} (2.36 \times 10^9 - 9.43 \times 10^9)$$

$$= 1.62 \times 10^{-18} \text{ J}$$

converting this to eV:

$$= \frac{1.62 \times 10^{-18}}{1.6 \times 10^{-19}}$$

$$= 10 \text{ eV}$$

Although the two values we have found are similar, it is clear that the value found by estimating the area under the graph is a slight overestimate.

### Potential in electric fields ($V_\text{E}$)

The potential difference between points **A** and **B** is defined as the work done per unit charge moving a charged body from **A** to **B**. (See Module 1 for the definition of potential difference.)

Since $\Delta W = \dfrac{Q_1 Q_2}{4\pi\varepsilon_o}\left(\dfrac{1}{r_1} - \dfrac{1}{r_2}\right)$

$$\Delta V_E = \dfrac{\Delta W}{Q_2} = \dfrac{Q_1}{4\pi\varepsilon_o}\left(\dfrac{1}{r_1} - \dfrac{1}{r_2}\right)$$

In this case, $\Delta V_E$ will be negative as

$$\dfrac{1}{r_2} > \dfrac{1}{r_1}$$

The absolute potential at a distance of $r$ from a charge $Q$ is given by:

$$V_E = \dfrac{Q}{4\pi\varepsilon_o r}$$

### Summary of field equations for electric fields

The equations for force, field strength, potential energy and potential for electric fields are all very similar. Figure 5.19 shows the relationships between the four equations. The following points may help you remember the equations and their interrelationship.

The arrows in Figure 5.19 indicate how to get from one field equation to another, i.e. they show the interrelationships between the equations.

- The two top equations relate to specific field situations. They are concerned with the interaction of *two* charged bodies. Therefore these equations have both charges mentioned in them.
- The lower two equations relate to properties of the field only. Therefore they mention only one charge — the one creating the field.
- The left hand equations concern force or field strength (which is defined in terms of force). These equations have $r^2$ in the fraction.
- The right hand equations concern energy or potential (which is defined in terms of energy). These equations have unsquared $r$ in the fraction.
- The parameters on the right are found by determining the area under the graph of the function on the left.
- The lower equations are produced by dividing the higher equation by the charge.
- The parameters on the left are found by determining the gradients of the graphs on the right. This fact is not required for your course, but is helpful to complete the overall picture.

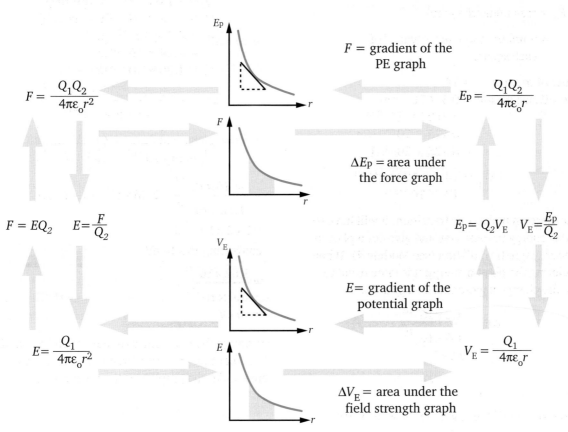

**Figure 5.19**

## Test your understanding

In these questions, $\varepsilon_o = 8.9 \times 10^{-12}$ Fm$^{-1}$.

1  An electric field is created by a point charge, $Q$, of $4.3 \times 10^{-8}$ C. Point **A** is 0.25 m from $Q$. Point **B** is 0.50 m from $Q$. Calculate:
   (a) the field strength at **A**
   (b) the potential at **A**
   (c) the field strength at **B**
   (d) the potential at **B**.

2  A $1.7 \times 10^{-10}$ C charge is placed at point **B** in the field described in (**1**). Calculate:
   (a) the force acting on the smaller charge
   (b) the work done moving the smaller charge to **B** from a position outside the field.

3  An atom is being ionised. The final electron starts at a distance of $2.4 \times 10^{-10}$ m from the nucleus. The charges on the electron and the nucleus are $1.6 \times 10^{-19}$ C and $8.0 \times 10^{-19}$ C, respectively.
   (a) Calculate the force between the electron and the nucleus.
   (b) By using the inverse square law, deduce values of force at distances of $4.8 \times 10^{-10}$ m, $9.6 \times 10^{-10}$ m and $19.2 \times 10^{-10}$ m.
   (c) Sketch a graph of force against distance for the values you have found.
   (d) By finding the appropriate area under the curve, find a value for the potential energy change of the electron as it is removed from the atom.
   (e) Use the appropriate formula to check your estimated value from (**d**).

## Radial Gravitational Fields

The analysis of gravitational fields is almost identical to the analysis of electric fields.

Fields created by a point mass are spherically symmetrical like the electric fields shown in Figures 5.11 and 5.12, although it is not usual to draw such diagrams for gravitational fields.

Fields created by spherical masses are identical to the fields created by point masses of the same mass for value of $r$ greater than or equal to the radius of the mass.

### Force and field strength in gravitational fields

Newton's Law of Gravitation states that the mutually attractive force experienced by two massive bodies is directly proportional to the product of their masses and inversely proportional to the square of their separation (see Figure 5.20).

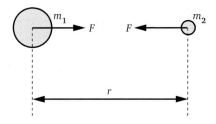

**Figure 5.20**

$$F \propto \frac{m_1 m_2}{r^2}$$

$$F = \frac{G m_1 m_2}{r^2}$$

$G$ is known as the universal gravitational constant and has a value of $6.7 \times 10^{-11}$ N m$^2$ kg$^{-2}$. Unlike the constant $\varepsilon$ in electric fields, which is dependant on the material, $G$ appears to have the same value in all situations.

Consider a field created by a large mass $M$. A smaller mass $m$ is placed into the field as shown in Figure 5.21.

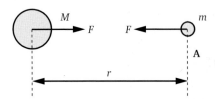

**Figure 5.21**

The force on $m$, caused by the field is given by:

$$F = \frac{G M m}{r^2}$$

The field strength at A, caused by $M$ is given by:

$$g = \text{force per unit mass} = \frac{F}{m}$$

$$g = \frac{F}{m} = \frac{GM}{r^2}$$

Figure 5.22 shows the variation of gravitational field strength with distance for the field around a point mass. Note that the variation is an inverse square law variation where doubling the distance decreases the field strength by a factor of four.

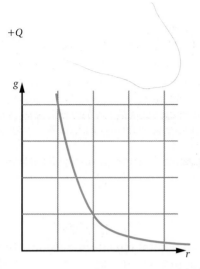

+Q

**Figure 5.22**

Figure 5.23 shows the variation of field strength with distance for a field in and around a solid spherical object of uniform density, e.g. a planet. Notice how the field strength is proportional to $r$ for values of $r$ less than the radius of the body. You will not be examined on this point but it may be instructive to discuss why the field strength varies in this way.

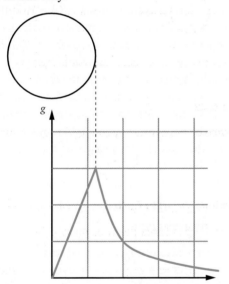

**Figure 5.23**

## Orbits of planets

It may be assumed that the orbits of planets around the Sun are circular. This also applies to the orbits of the Moon and artificial satellites around the Earth (see Figure 5.24).

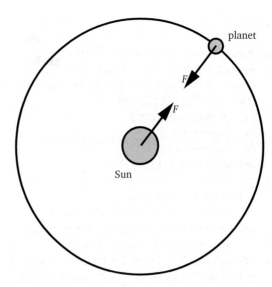

**Figure 5.24**

The Sun and the planets attract each other. The force of attraction $F$ provides the centripetal force which causes circular motion. Both the Sun and the planet will orbit about a point that is the centre of gravity of the system. However, since suns are much, much more massive than planets, it is safe to assume that the centre of gravity of the combined Sun and planet is the same as the centre of gravity of the Sun. We shall consider that the Sun stays still while the planet orbits around it.

For the orbit shown in Figure 5.24:

$$F = \frac{GM_s M_p}{r^2}$$

Since the gravitational force is also the centripetal force:

$$F = \frac{M_p v^2}{r}$$

where $v$ is the speed of the planet. See Unit 4 for details of circular motion.

Equating these two expressions for $F$ gives:

$$\frac{GM_s M_p}{r^2} = \frac{M_p v^2}{r}$$

Some of the terms cancel out, leaving:

$$\frac{GM_s}{r} = v^2$$

or:

$$v = \sqrt{\frac{GM_s}{r}}$$

Notice that:
- the orbital speed of a planet is independent of its mass, but it does depend on the mass of the Sun
- the orbital speed will be less for planets with greater orbital radii.

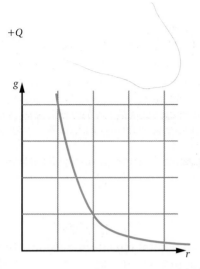

In order to find the orbital period for a planet, it is more convenient to equate the gravitational attraction to its alternative version of the expression for centripetal force:

$$\frac{GM_s M_p}{r^2} = M_p r \omega^2$$

where $\omega$ is the angular speed of the planet.

Rearranging this equation gives:

$$\omega^2 = \frac{GM_s}{r^3}$$

since:

$$\omega = \frac{2\pi}{T}$$

$$\frac{4\pi^2}{T^2} = \frac{GM_s}{r^3}$$

or:

$$T^2 = \frac{4\pi^2 r^3}{GM_s}$$

Notice again that the orbital period is independent of the planet's own mass and depends only on the Sun's mass and on the radius of the orbit.

## Solar system data

The following data need not be learned but will be used in numerical questions in this chapter. When performing calculations involving orbits, remember to:

- convert all masses to kg
- give all times in seconds, even if they are quoted in hours.

**Table 5.4**

| Body | Mass/kg | Orbital radius/m |
|------|---------|------------------|
| Sun | $2.0 \times 10^{30}$ | – |
| Mercury | $3.3 \times 10^{23}$ | $5.8 \times 10^{10}$ |
| Venus | $4.9 \times 10^{24}$ | $1.1 \times 10^{11}$ |
| Earth | $6.0 \times 10^{24}$ | $1.5 \times 10^{11}$ |
| Mars | $6.4 \times 10^{23}$ | $2.3 \times 10^{11}$ |
| Jupiter | $1.9 \times 10^{27}$ | $7.8 \times 10^{11}$ |
| Saturn | $5.7 \times 10^{26}$ | $14.3 \times 10^{11}$ |
| Uranus | $8.7 \times 10^{25}$ | $28.7 \times 10^{11}$ |
| Neptune | $1.0 \times 10^{26}$ | $45.0 \times 10^{11}$ |
| Pluto | $1.3 \times 10^{22}$ | $59.1 \times 10^{11}$ |

## Orbits of the Earth's Satellites

Earth has a natural satellite, the Moon. The Moon has a mass of $7.4 \times 10^{22}$ kg.

The analysis of the Moon's orbit around the Earth is almost identical to the forgoing analysis of the planets' orbits around the Sun. The key starting point is to equate the centripetal force causing the orbit to the gravitational attraction between the two bodies.

The Earth also has many artificial satellites, the orbits of which may be analysed in the same way.

Satellites used for communication (see Module 2) have to be in a fixed position relative to the Earth's surface. Otherwise, in order to receive a "satellite" signal for your television you would have to continually adjust the direction in which your satellite aerial was pointing. As the Earth rotates, so must the satellite. Such a satellite must be placed:

- above the equator
- in an orbit that rotates in the same direction in which the Earth spins on its axis
- in an orbit that has a period of 24 hours.

An orbit that fulfils these requirements is called a **geosynchronous** orbit.

Satellites are also used to collect data for meteorological and geological surveys. For these and other purposes, frequent overflying by the satellite is needed. Figure 5.25 shows a satellite in a polar orbit.

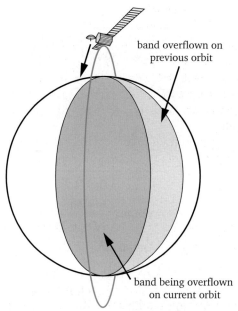

band overflown on previous orbit

band being overflown on current orbit

**Figure 5.25**

If one orbit takes 2 hours to complete, each time it flies over, the satellite can photograph a band of the Earth's surface. Between one orbit and the next, the Earth will have rotated by an angle of 30°, so that a different band of the Earth's surface is viewed on each orbit. There are several advantages to this type of satellite:

- they make frequent passes over any given place on the Earth's surface, so that the way in which changes develop can be monitored
- to make the orbital period small, the orbital radius has to be small. The satellites are orbiting quite close to the Earth's surface, just above the atmosphere, and can pick out fine detail.

**Test your understanding**

In these questions, G = 6.7 × 10⁻¹¹ Nm² kg⁻².
1 Taking the period of the Moon's orbit to be 28 days, calculate a value for the orbital radius.
2 Use your value for the orbital radius to calculate a value for the strength of the Earth's gravitational field at the Moon.
3 Calculate a value for the strength of the Moon's gravitational field at its surface.
(The radius of the moon is $1.7 \times 10^6$ m.)
4 A body is placed between the Moon and the Earth in a position where it feels equal attraction to both bodies. How far is it from the Earth?
(The radius of the Moon's orbit is $3.8 \times 10^8$ m.)
5 For a satellite in a geosynchronous orbit, calculate:
(a) the orbital radius
(b) the orbital height above the Earth's surface.
(The radius of the Earth is $6.4 \times 10^6$ m.)
6 Repeat question **5** for a satellite having an orbital period of 2.0 hours.

## Changes of Potential Energy in Gravitational Fields

Satellites launched from the Space Shuttle and destined for a geosynchronous orbit are first put into an orbit about 300 km above the Earth's surface and then 'boosted' into their final orbit. As the height of their orbit is increased, their potential energy is also increased as work must be done against gravitational attraction.

However, to be able to calculate the change in potential energy we cannot simply use the equation $\Delta E_p = mg\,\Delta h$, because the gravitational field strength, $g$, will vary as the satellite is lifted. To make a proper calculation of the change in potential energy it is necessary to perform an analysis similar to that for potential energy changes in electric fields.

Consider a mass $m_1$ 'creating' a gravitational field (see Figure 5.26). Consider a small mass $m_2$ placed an infinite distance from $m_1$. Since the two masses cannot influence each other, the potential energy stored in the field is zero. When $m_2$ is moved to a position, **P**, closer to $m_1$, its potential energy will change. The potential energy at **P** will be equal to the amount of work that must be done moving $m_2$ from infinity to **P**.

However, $m_1$ and $m_2$ are attracting each other, so no positive work needs to be done to move $m_2$ to **P**: work would have to be done in order to move $m_2$ from **P** to infinity. In other words, the potential energy when $m_2$ is at **P** must be *less* than when $m_2$ is at infinity. Since the potential energy when $m_2$ is at infinity is zero, the potential energy at **P** must be negative.

Figure 5.26 shows the satellite mentioned above in its two positions having orbital radii of $r_1$ and $r_2$ respectively. The figure also shows a graph of variation of force, $F$, with orbital radius.

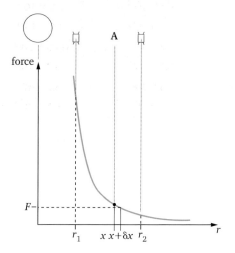

**Figure 5.26**

As the satellite is moved from its orbit of radius $r_1$, to its orbit of radius $r_2$, it passes the point **A** (orbital radius $x$). At **A**, the satellite is moved the small additional distance $\delta x$, which is so small that the force $F$ can be considered to be constant over this distance. The work done in moving the satellite the distance $\delta x$ is $\delta W$.

$$\delta W = F\delta x$$

This is equivalent to the area of the rectangle under the graph at **A**. The 'boost' from one orbit to the other can be considered to be made up of a very large number of small movements, each of distance $\delta x$. The work done, $\delta W$, for each of these movements will be different, as $F$ is different for each small movement. Nevertheless, each $\delta W$ is equivalent to the area of the appropriate rectangular strip under the graph (see Figure 5.27).

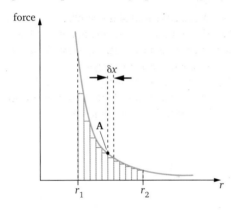

**Figure 5.27**

It may be seen that the total work done, $W$, is the sum of all the values $\delta W$. Therefore $W$ is equivalent to the area under the graph between $r_1$ and $r_2$. As the satellite changes its orbit, its change in potential energy is also equivalent to the area under the force–distance graph.

At **A**:

$$F = \frac{Gm_1m_2}{x^2}$$

$$\delta W = \frac{Gm_1m_2}{x^2}.\delta x$$

$$W = \sum \frac{Gm_1m_2\delta x}{x^2}$$

This summation can be done using calculus and will produce the result:

$$W = Gm_1m_2\left(\frac{1}{r_1} - \frac{1}{r_2}\right)$$

So:     $$\Delta E_p = Gm_1m_2\left(\frac{1}{r_1} - \frac{1}{r_2}\right)$$

This equation is valid for any values of $r_1$ and $r_2$.

## When to use $\Delta E_p = mg\Delta h$

Consider a body of mass $m_2$ lifted a height $\Delta h$ from the surface of the Earth of mass $m_1$ and radius $r_1$.

$$\Delta E_p = Gm_1m_2\left(\frac{1}{r_1} - \frac{1}{r_2}\right)$$

but:     $$r_2 = r_1 + \Delta h$$

so:     $$\Delta E_p = Gm_1m_2\left(\frac{1}{r_1} - \frac{1}{r_1 + \Delta h}\right)$$

$$= Gm_1m_2\left(\frac{r_1 + \Delta h - r_1}{r_1(r_1 + \Delta h)}\right)$$

$$= \frac{Gm_1m_2\Delta h}{r_1^2 + r_1\Delta h}$$

If $\Delta h$ is very small compared with $r_1$:

$$r_1^2 + r_1\Delta h \approx r_1^2$$

$$\Delta E_p \approx m_2\frac{Gm_1}{r_1^2}.\Delta h$$

but $\dfrac{Gm_1}{r_1^2} = g = $ gravitational field strength

$$\therefore \Delta E_p \approx m_2g\Delta h$$

As you can see, the simple equation for change in potential energy is only valid when $\Delta h$ is very small compared with the radius of the Earth.

## The potential energy/separation graph

So far, we have looked at the change in potential energy when the separation of two bodies changes from $r_1$ to $r_2$. The actual potential energy when $m_2$ is at a distance of $r$ from $m_1$ is given by:

$$E_p = \frac{-Gm_1m_2}{r}$$

Remember that potential energy when $r$ is infinite is zero. As described above, potential energy decreases as $r$ gets smaller, so that all values of potential energy in gravitational fields are negative. This is a consequence of all gravitational forces being attractive. Figure 5.28 shows the variation of gravitational potential energy with separation, $r$.

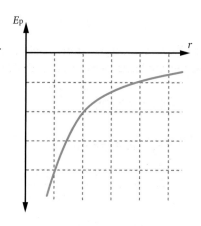

**Figure 5.28**

Notice that the magnitude of potential energy halves as $r$ doubles.

## Potential in Gravitational Fields

The potential at a point is defined as the work done *per unit mass* in moving a body from infinity to that point. As the gravitational field is always attractive, work would have to be done in moving a mass from the point to infinity. Potential, like potential energy, is always negative in gravitational fields.

For a gravitational field 'created' by the mass $m_1$, with a smaller mass $m_2$ placed at a distance $r$ from $m_1$:

$$E_p = \frac{-Gm_1m_2}{r}$$

then:

$$V_G = \frac{E_p}{m_2} = \frac{-Gm_1}{r}$$

### Test your understanding

1  Calculate the potential in the Earth's gravitational field at the position of:
   (a) a geosynchronous orbit;
   (b) an orbit having a period of 2.0 h.

2  Calculate the absolute potential energy of a 450 kg satellite:
   (a) at the surface of the Earth (of radius $6.4 \times 10^6$ m);
   (b) when it is in an orbit 400 km above the surface of the Earth.

3  A 200 kg satellite changes orbit from one with a radius of $8.1 \times 10^6$ m to one of radius $4.2 \times 10^7$ m. Calculate the change in potential energy.

4  A 320 kg satellite is lifted from an orbit $1.2 \times 10^6$ m above the surface of the Earth to one which is $3.6 \times 10^7$ m above the Earth's surface. Calculate:
   (a) the initial speed of the satellite;
   (b) the initial kinetic energy of the satellite;
   (c) the change in potential energy when the satellite's orbit is changed.
   (d) the new kinetic energy of the satellite.
   (e) the overall change in the sum of the satellite's kinetic and potential energies.

## Summary of field equations for gravitational fields

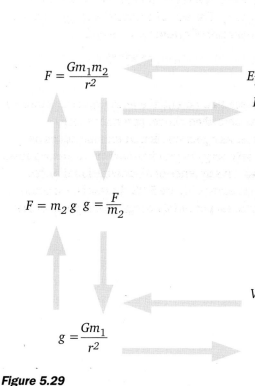

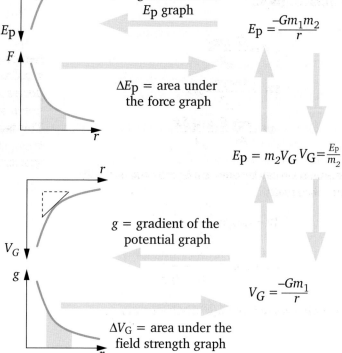

$$F = \frac{Gm_1m_2}{r^2}$$

$F$ = gradient of the $E_p$ graph

$$E_p = \frac{-Gm_1m_2}{r}$$

$\Delta E_p$ = area under the force graph

$$F = m_2 g \quad g = \frac{F}{m_2}$$

$$E_p = m_2 V_G \quad V_G = \frac{E_p}{m_2}$$

$g$ = gradient of the potential graph

$$g = \frac{Gm_1}{r^2}$$

$$V_G = \frac{-Gm_1}{r}$$

$\Delta V_G$ = area under the field strength graph

**Figure 5.29**

All four of these main relationships should be learned, as should the equations and processes connecting them. Notice that the potential energy at a point in a field is equal to the gradient of the force graph at that point. Similarly, the potential is the same as the gradient of the field strength graph.

### Test your understanding

**1** The mass of the Earth is $6.0 \times 10^{24}$ kg and the radius is $6.4 \times 10^{6}$ m. The universal gravitational constant $G = 6.7 \times 10^{-11}$ N m$^2$ kg$^{-2}$. A satellite of mass 260 kg is lifted into an orbit of radius $4.0 \times 10^{7}$ m.
   **(a)** Plot a graph of the variation of field strength with orbital radius over the range 0 to $4.0 \times 10^{7}$ m.
   **(b)** Use your graph to determine the change in potential between the Earth's surface and satellite's orbital radius.
   **(c)** Calculate the satellite's change in potential energy when it is lifted into its orbit.

**2** **(a)** Plot a graph of field strength against separation for the Earth's gravitational field. Your graph should cover a range of separations from $6.4 \times 10^{6}$ m to $4.0 \times 10^{7}$ m.
   **(b)** Use your graph to find the gravitational field strength at a distance from the centre of the Earth of $2.0 \times 10^{7}$ m.
   **(c)** Check your answer to **(b)** by making a direct calculation using the appropriate formula.

**3** A body of mass 160 kg is lifted from the Earth's surface. For each of the following examples, calculate:
   **(i)** the change in potential energy using the formula $\Delta E_p = mg\Delta h$;
   **(ii)** the change in potential energy using the formula:

$$\Delta E_p = Gm_1 m_2 \left( \frac{1}{r_1} - \frac{1}{r_2} \right).$$

   **(iii)** the percentage error involved in using the approximate formula.
   (Take $g$ to be 9.814 N kg$^{-1}$)
      **(a)** lifting it onto a table of height 0.80 m;
      **(b)** lifting it onto the roof of an office building of height 63 m;
      **(c)** lifting it to the summit of Scafell Pike (height 978 m);
      **(d)** lifting it into an orbit of height 300 km.

**4** A gravitational field is created by a point mass. The field strength at a distance of $2.0 \times 10^{11}$ m from the mass is $4.0 \times 10^{-9}$ N kg$^{-1}$.
   **(a)** What will be the field strength at distances of $4.0 \times 10^{11}$ m and $1.0 \times 10^{12}$ m?
   **(b)** The potential of the field at a distance of $2.0 \times 10^{11}$ m is $-8.0 \times 10^{-2}$ J kg$^{-1}$. What will be the potentials at distances of $4.0 \times 10^{11}$ m and $1.0 \times 10^{12}$ m?

### Similarities between gravitational and electric fields

From Figures 5.19 and 5.29, you can see that for both fields:

$$\text{force} \propto \frac{1}{r^2}$$

$$\text{field strength} \propto \frac{1}{r^2}$$

$$\text{potential energy} \propto \frac{1}{r}$$

$$\text{potential} \propto \frac{1}{r}$$

Also for both fields:

$$\text{force} \propto \text{ product of masses or product of charges}$$

$$\text{potential energy} \propto \text{ product of masses or product of charges}$$

and

$$\text{field strength} \propto \text{ mass or charge 'creating' the field}$$

$$\text{potential} \propto \text{ mass or charge 'creating' the field}$$

### Differences between gravitational and electric fields

- Gravitational fields are always attractive. Forces in electric fields can attract or repel.
- Gravitational potential and potential energy are therefore always negative. Electric fields can have positive or negative values for potential and potential energy. In fact, since potential energy is defined as the work done moving unit *positive* charge to the point in the field, the equation for potential seems to imply that potential is positive. Of course, where the charge causing the field is negative the value for potential will also be negative.
- The nature of the medium for electric fields affects the force, field strength potential and potential energy as different media will have different values of $\varepsilon$. However $G$ seems to be a universal constant, so the medium does not affect gravitational fields.
- Electric fields tend to be important on a small scale, e.g. they play a significant role in atoms and even nuclei. Gravitational fields tend to have greater significance on a larger scale, e.g. with respect to planets, galaxies, etc.

# Gravitation and the Future of the Universe

The universe is expanding and has been doing since the Big Bang. It is not known whether this expansion will continue forever.

Gravitational forces are attracting all the material in the universe back together. If these gravitational forces are strong enough they will slow the expansion, stop it altogether and bring all the material back together again. Galaxies would collide and become one super galaxy. Matter would be forced closer and closer together, even collapsing atoms. This idea is known as the *closed* model of the universe. It will only happen if the universe's gravitational field strength is sufficient to reverse the expansion. Since this field strength is proportional to mass, it is clear that the universe will be 'closed' if it has sufficient mass.

The opposite scenario is that there is insufficient mass in the universe to cause its expansion to be reversed. If this is the case, the expansion may slow down, but will continue permanently. This is the so-called *open* model of the universe. Were this to happen, energy and matter would spread uniformly through the universe producing a cool uniformity.

With a mass somewhere in between these two extremes, the universe could continue to expand but at an ever-slowing rate. The universe would then be said to have **critical density**.

Most of the current evidence suggests that there is insufficient mass in the universe to halt the expansion. There are two areas of uncertainty:

- some of the universe's matter is *dark matter*, which is hard to detect as it gives off no radiation.
- it is not certain whether neutrinos (Module 2) have zero mass or simply very little mass. If they have finite mass this will greatly increase the known mass of the universe since there are so many of them.

**Test your understanding**

1  In an experiment (first done by Milikan) to determine the charge on an electron, a charged oil drop is held stationary between two parallel plates where a uniform electric field exists. The mass of the oil drop has been found.

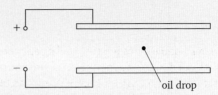

**Figure 5.30**

(a) (i) Explain what is meant by a uniform electric field.

   (ii) Sketch a diagram to show the electric field between and at the edges of the plate.

(b) In one test, an oil drop has a weight of $3.0 \times 10^{-14}$ N. The plates are 4.0 mm apart and the potential difference between them is 380 V.

   (i) Calculate the field strength between the plates.

   (ii) Calculate the magnitude of the charge on the oil drop.

   (iii) State and explain whether the charge is positive or negative.

(c) A second, unobserved oil drop, which has an equal but opposite charge, is 0.10 mm from the first drop. ($\varepsilon_o = 8.9 \times 10^{-12}$ F m$^{-1}$)

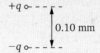

**Figure 5.31**

Show that the magnitude of the force between the two drops may be neglected in the calculations in (b).

**2** The graph shows the variation of electric field strength, $E$, with distance, $r$, from the centre of a hydrogen nucleus.

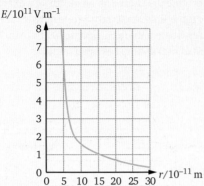

**Figure 5.32**

(a) (i) Determine the value of the electric field strength at a distance of $5.0 \times 10^{-11}$ m from the centre of the nucleus.

(ii) Use your value of the electric field strength to calculate the magnitude of the force between the nucleus and an electron separated by a distance of $5.0 \times 10^{-11}$ m. ($e = 1.6 \times 10^{-19}$ C.)

(b) Briefly explain how the graph could be used to estimate the amount of work required to remove the electron from the atom, starting with the electron at a distance of $5.0 \times 10^{-11}$ m from the nucleus.

(c) Calculate the magnitude of the gravitational force between the electron and the proton at a distance of $5.0 \times 10^{-11}$ m and show that gravitational attraction does not contribute significantly to keeping the two together.
($G = 6.7 \times 10^{-11}$ N m² kg⁻²,
mass of proton $= 1.7 \times 10^{-27}$ kg,
mass of electron $= 9.1 \times 10^{-31}$ kg.)

**3** (a) (i) State a formula for the gravitational force between a satellite and the Earth given that:

R is the radius of the orbit
G is the universal gravitational constant
m is the mass of the satellite
M is the mass of the Earth.

(ii) By equating the gravitational force with the centripetal force, show that the period, $T$, of a satellite's orbit is given by:

$$T^2 = \frac{4\pi^2 R^3}{GM}$$

(b) Calculate the radius of the orbit of a satellite that has an orbital period of 2.0 hour.
$G = 6.7 \times 10^{-11}$ N m² kg⁻²,
$M = 6.0 \times 10^{24}$ kg.

(c) Explain briefly why it requires more energy to lift a satellite into a geosynchronous orbit than it takes to lift a similar satellite into an orbit having a period of 2.0 hours.

# Magnetic Fields

## Magnetic Fields

A magnetic field is a region in which any of the following will experience a force:

- a permanent magnet
- a piece of a magnetic material
- a conductor carrying an electric current
- a moving charged particle.

Permanent magnetism is not well understood, and most of this section will deal with magnetism caused by and affecting electricity.

### Drawing magnetic fields

Figure 5.33 shows the familiar magnetic field around a bar magnet.

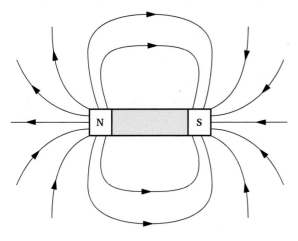

**Figure 5.33**

Magnetic field lines are the paths that would be followed by isolated North magnetic poles, were they free to move. Note the following concerning magnetic field lines illustrating magnetic fields:

- the field lines are continuous and unbroken
- the field lines do not touch or cross
- where the field lines are close together, the magnetic field strength (otherwise known as **magnetic flux density**, $B$) is greatest
- the magnetic field lines have direction, from North to South, indicated by arrows. This correctly suggests that magnetic flux density is a vector quantity (see Module 1).

Figure 5.34 shows the magnetic field between two ceramic magnets.

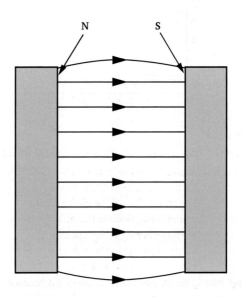

**Figure 5.34**

Notice that the field is uniform over most of its extent: the field lines are all in the same direction and are the same distance apart. It is only at the edges that the field is distorted.

You are not required to learn these magnetic field patterns, but it is useful to know how to draw the fields.

It is often necessary to represent fields perpendicular to the plane of the paper. This is shown in figure 5.35.

⊗ this represents a magnetic field directed into the paper

⊙ this represents a magnetic field directed out of the paper

**Figure 5.35**

## Magnetic Flux and Flux Density

Some magnetic fields are stronger than others. The quantity you are assessing when you make a judgement about the strength of a magnet is the flux density. A reasonable way of visualising a strong magnetic field is to picture a field with magnetic field lines that are packed close together. The flux density may be thought of as the number of flux lines that pass through each square metre. This is *not* a definition of flux density however.

The symbol for flux density is $B$ and its unit is the tesla (T).

The flux in a magnetic field may be visualised as the total number of magnetic field lines rather than their concentration. Once again, be aware that this is *not* a definition.

The symbol for flux is $\phi$ (the Greek letter phi) and its unit is the weber (Wb).

Clearly, in any situation, these two quantities are linked by the equation:

$$\phi = BA \quad \text{where } A = \text{area}$$

or:

$$B = \frac{\phi}{A}$$

The units of flux and flux density are linked as follows:

$$1 \text{ T} = 1 \text{ Wb m}^{-2}$$

## Forces on Current Carrying Conductors

A current carrying wire placed at right angles to a magnetic field will experience a force. The force will be at right angles to both the field and the current (see Figure 5.36).

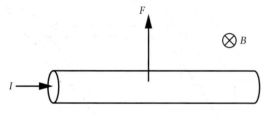

**Figure 5.36**

The direction of the force can be found by using Fleming's Left Hand Rule as illustrated below in Figure 5.37.

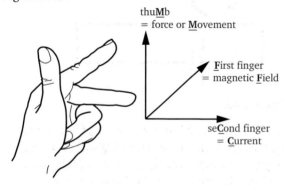

thu**M**b
= force or **M**ovement

**F**irst finger
= magnetic **F**ield

se**C**ond finger
= **C**urrent

**Figure 5.37**

Clearly, the magnitude of the force, $F$, experienced by the conductor will depend on the magnetic flux density, the current and the length of the conductor.

It is not surprising that:

$$F \propto BIl$$

or:

$$F = k.BIl$$

where $k$ is a constant.

By choosing the unit of magnetic flux density (the tesla) to be the appropriate size, the constant ($k$) is made to be 1. So:

$$F = BIl$$

The tesla can be defined as that magnetic flux density that will lead to the production of a force of 1 N on a current of 1 A flowing in a wire of length 1 m placed at right angles to the magnetic field.

This equation is only valid if the conductor is at right angles to the magnetic field. Figure 5.38 shows the way in which the force varies with the angle, $\theta$, between the current and the magnetic flux density.

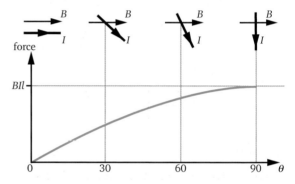

**Figure 5.38**

It is not necessary to learn this variation for examination purposes.

## The Current Balance

Current balances are devices that can be used to measure the strengths of magnetic fields. The copper wire in Figure 5.39 carries a current, $I$, through the magnetic field, $B$, provided by the magnets.

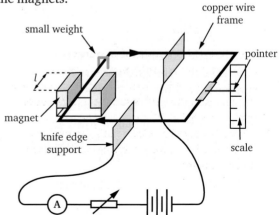

**Figure 5.39**

Before the current is turned on, the wire is balanced. When the current is turned on, the force, F, experienced by the wire is in an upward direction. Additional small weights of mass, $m$, must be added to the frame in order to re-balance it. Since the weights are supplying a force that is equal and opposite to F:

$$BIl = mg$$

Provided that $m$, $l$ and $I$ are known, the magnetic flux density, $B$, can be determined.

The balance can be adapted to measure the force between parallel wires, each carrying an electric current (see Figure 5.40).

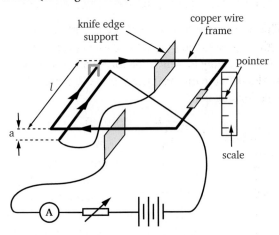

**Figure 5.40**

The amp is defined in terms of the force that is experienced between two parallel conductors carrying electric currents.

**Test your understanding**

1 Complete the table below.

**Table 5.5**

| | B | A | $\phi$/Wb |
|---|---|---|---|
| a | 0.47 T | $2.4 \times 10^{-4}$ m$^2$ | |
| b | | 6.2 cm$^2$ | $8.6 \times 10^{-5}$ |
| c | 16 mT | | 0.052 |

2 The average flux density of the vertical component of the Earth's magnetic field is approximately $4.5 \times 10^{-5}$ T. Estimate the value of the flux, $\phi$, at the Earth's surface. The radius of the Earth is $6.4 \times 10^6$ m.

3 Complete the table below relating to the forces experienced by current carrying conductors in magnetic fields.

**Table 5.6**

| | B/T | F/N | I | I |
|---|---|---|---|---|
| a | 0.24 | | $6.2 \times 10^{-2}$ m | 3.6 A |
| b | | $2.6 \times 10^{-3}$ | 24 m | $1.6 \times 10^{-2}$ A |
| c | 2.8 | 0.067 | | 42 mA |
| d | $5.0 \times 10^{-5}$ | $9.2 \times 10^{-6}$ | 82 cm | |

4 In a current balance like the one shown in Figure 5.39 (page 104), the frame is rebalanced using a mass of 0.63 g. The current is 4.2 A and 5.6 cm of the frame is affected by the uniform magnetic field. The area of the magnets is $1.4 \times 10^{-3}$ m$^2$ and $g$ is 9.8 N kg$^{-1}$. Calculate:
 (a) the flux density of the magnetic field between the magnets
 (b) the flux between the magnets.

5 If the magnets in (4) were angled so that the field was at 45° to the wire, what would be the magnitude of the force on the wire? (Hint: find the component of the magnetic field at right angles to the wire: see Module 1.)

## The Simple Motor

Figure 5.41 is a simplified diagram of the main features of an electric motor.

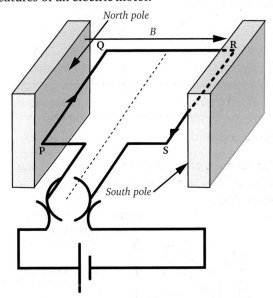

**Figure 5.41**

Use Fleming's Left Hand Rule on the conductors **PQ** and **RS**. It can be seen that **PQ** will be forced downwards and, since the current in **RS** is in the opposite direction, it will be forced upwards. The rotor will tend to spin anticlockwise.

When the rotor has turned through an angle of 180°, **PQ** will have moved round so that it forms the right-hand side of the rotor and **RS** will now be on the left (see Figure 5.42).

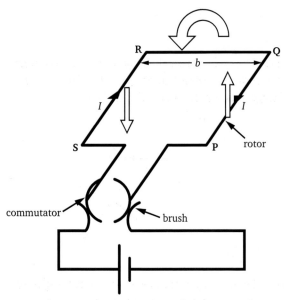

**Figure 5.42**

The brushes and commutator ensure that the current still enters the rotor on the left-hand side. Although the positions of the conductors in the rotor have reversed, the part of the rotor that is on the left is still being forced down and the side on the right is being forced up. The rotor continues to rotate in an anticlockwise direction.

### The variation of torque with rotor position

The rotor will experience its maximum torque when it is in the position shown in Figure 5.41. Assuming that the magnetic field is uniform, the magnitude of the forces experienced by each side of the rotor, $F$, will not change as the rotor turns. However, the perpendicular distances between the forces, $F$, and the axis of the rotor *will* change and so will the moment or torque experienced by the rotor (see Module 1).

When the normal to the rotor makes an angle of 90° with the magnetic field, each force, $F$, is a distance of $\dfrac{b}{2}$ from the axis of the rotor (see Figure 5.43).

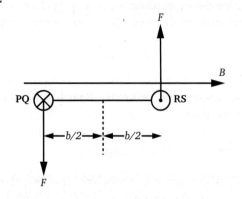

**Figure 5.43**

The total torque on the rotor is given by:

$$\text{torque} = \text{torque on } \mathbf{PQ} + \text{torque on } \mathbf{RS}$$
$$= F.\frac{b}{2} + F.\frac{b}{2}$$
$$= Fb$$

When the normal to the rotor makes an angle $\theta$ with the magnetic field (see Figure 5.44), the torque is given by:

$$\text{Torque} = F\,b\,\sin\theta$$

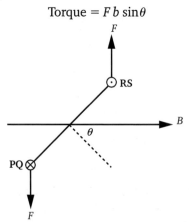

**Figure 5.44**

When $\theta$ is 0° and the rotor is at right angles to the magnetic field there is zero torque. In any case, at this point, the small gap in the commutator would mean that there is no current in the rotor. The rotor is not being forced round at this point: it is merely its inertia that carries it past this point to a position where a torque begins to establish itself once more. You are not required to memorise the formulae for the dependence of torque on rotor angle. However, you may be expected to deal with questions that combine ideas from this section of the specification with other ideas such as the work on moments from Module 1.

## Forces on Moving Charges

Figure 5.45 shows a current carrying conductor in a magnetic field.

The current is the movement of free electrons through the wire (see Module 1). Some of the electrons are shown schematically in Figure 5.45. Each electron that is moving at right angles to the magnetic field experiences a small force, $f$.

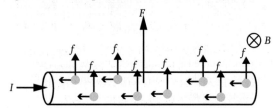

**Figure 5.45**

The total force on the wire, $F$, is made up of the sum of all the small forces on the individual electrons.

$$F = Nf$$

where $N$ is the total number of free electrons moving through the wire.

$$N = nAl$$

where $n$ is the number of free electrons per unit volume, $A$ is the cross-sectional area of the wire and $l$ is its length (see Module 1).

also: $\qquad\qquad F = BIl$

so: $\qquad\qquad BIl = nAlf$

or: $\qquad\qquad BI = nAf$

but since: $\qquad\qquad I = nAve$

where $v$ is the mean electron drift velocity and $e$ is the charge on an electron:

$$BnAve = nAf$$

or: $\qquad\qquad f = Bev$

On your formulae sheet, $F$ is used instead of $f$. In general, for a particle of charge $q$, rather than the special case of an electron:

$$F = Bqv$$

**Test your understanding**

1. A motor like that in Figure 5.41 (page 105) has a rotor with a length (PQ) of 6.0 cm and a breadth of 4.0 cm. The current in the rotor is 1.8 A and the flux density of the magnetic field is 0.089 T. Calculate:
   (a) the force experienced by the wire PQ
   (b) the maximum torque experienced by the rotor.

2. In a motor in which the magnetic field is from right to left and the current is flowing anticlockwise (viewed from above), determine the direction of rotation of the rotor.

3. Complete the following table relating to the forces experienced by charged particles moving in magnetic fields.

**Table 5.7**

|   | B/T | q/C | F/N | v/m s$^{-1}$ |
|---|-----|-----|-----|--------------|
| a | 2.4 | $1.6 \times 10^{-19}$ | | $1.5 \times 10^{8}$ |
| b | 0.36 | $3.2 \times 10^{-19}$ | $6.7 \times 10^{-14}$ | |
| c | $8.2 \times 10^{-3}$ | | $3.4 \times 10^{-13}$ | $6.5 \times 10^{7}$ |
| d | | $4.8 \times 10^{-18}$ | $2.2 \times 10^{-12}$ | $8.8 \times 10^{6}$ |

4. An electron (of charge $1.6 \times 10^{-19}$ C and mass $9.1 \times 10^{-31}$ kg) is accelerated by a potential difference of 3400 V. It is then subjected to a magnetic field of flux density 0.056 T. Calculate:
   (a) the speed of the electron
   (b) the force it experiences when it enters the magnetic field.

# The Movement of Charged Particles in Magnetic Fields

Figure 5.46 shows an electron moving in a uniform magnetic field of flux density $B$. The electron, of charge $e$, enters the field with a velocity $v$.

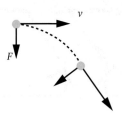

**Figure 5.46**

The electron experiences a force at right angles to both the field and to its instantaneous velocity. Notice that, when using Fleming's Left Hand Rule to judge the direction of the force, the second (current) finger should point in the opposite direction to the electron's velocity. This is because the conventional current is opposite to the direction of motion of the negatively charged electrons (see Module 1).

- The electron experiences no acceleration in the direction of its existing velocity as the force is at right angles to this direction.
- The electron does experience an acceleration at right angles to its existing velocity.
- When the electron has changed direction, the direction of the force that it experiences will also change: the force will always be at right angles to its current velocity.
- The magnitude of the new force is the same as the old force since the electron has not changed its speed.
- In other words, the electron will travel in a circular path. The centripetal force that causes the circular motion is $F (= Bev)$.

The radius, $r$, of the circular path can be found by equating the expressions for centripetal force and for the force on a moving charge in a magnetic field:

$$Bev = \frac{mv^2}{r}$$

where $m$ is the electron's mass. Rearranging this equation gives:

$$r = \frac{mv}{eB}$$

Details of instruments in which the movements of charged particles are affected by magnetic fields are given on page 126 *et seq.* In most of these machines, it is required to select particles of a single velocity.

## Velocity Selectors

Figure 5.47 shows an electron moving through a region in which there is a magnetic field and an electric field.

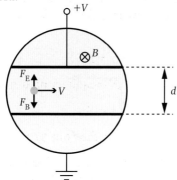

**Figure 5.47**

The electron experiences a force, $F_E$, due to the electric field and a force, $F_B$, due to its movement through the magnetic field. The equations for these forces are:

$$F_E = \frac{eV}{d} \qquad F_B = Bev$$

where:  $e$ = charge on electron
 $V$ = p.d. between plates
 $d$ = plate separation
 $v$ = electron speed
 $B$ = flux density.

If the two forces are exactly equal and opposite, there will be no resultant force on the electron (see Module 1) and it will pass through the region undeviated.

$$F_E = F_B$$

$$\frac{eV}{d} = Bev$$

or:  $$v = \frac{V}{Bd}$$

Only electrons of velocity $v$ will follow a straight path through the combined fields. The paths of electrons with other initial velocities are shown in Figure 5.48.

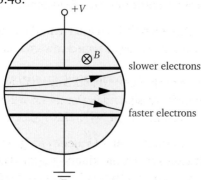

slower electrons

faster electrons

**Figure 5.48**

1 Explain why the slower and faster electrons in Figure 5.48 follow the paths shown.

2 An electron ($e = 1.6 \times 10^{-19}$ C, $m = 9.1 \times 10^{-31}$ kg) with a speed of $4.6 \times 10^7$ m s$^{-1}$, enters a magnetic field ($B = 1.3$ mT). Calculate:
   (a) the magnitude of the force acting on the electron;
   (b) the radius of its path.

3 Explain why an electron entering a magnetic field at right angles to its velocity follows a circular path.

4 (a) A positive ion of mass $3.4 \times 10^{-25}$ kg and charge $3.2 \times 10^{-19}$ C is to travel in a circular path of radius 0.50 m. The speed of the ion is $3.6 \times 10^5$ m s$^{-1}$. Calculate the flux density of the magnetic field that would be required.
   (b) Sketch a diagram showing the relative directions of the magnetic field and the path of the ion.

5 An electron ($e = 1.6 \times 10^{-19}$ C) enters a velocity selector. The potential difference across the plates is 2.0 kV and they are separated by a distance of 4.0 cm. The magnetic field has a flux density of 64 mT.
   (a) Write out an expression for the force acting due to the electric field. Define your terms.
   (b) Write out an expression for the force acting due to the magnetic field. Define your terms.
   (c) Calculate the speed of the electrons which pass through the selector in a straight line.

## Electromagnetic Induction

When the magnetic field around a conductor changes, a voltage or *emf* is produced in the conductor. The word *induced* is usually used instead of *produced*, so the voltage is usually called an induced emf.

Figure 5.49 shows a magnet being moved into a coil.

## Qualitative demonstration of induced emfs

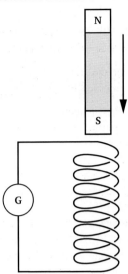

**Figure 5.49**

The coil is connected to a sensitive voltmeter or galvanometer. When the magnet is moved into or out of the coil, the size and direction of the induced emfs are noted on the galvanometer. Empirical results and deductions are as follows.

- An emf is only induced when the magnet is moving.
- **An emf is induced when the magnetic field around a conductor changes (Faraday's 1st law of electromagnetic induction).**
- The size of the emf depends on the speed of the magnet: faster movements produce greater emfs.
- The size of the emf depends on the number of turns on the coil: if there are more turns, a greater emf is produced.
- The size of the induced emf depends on the strength of the magnets: stronger magnets produce greater emfs.
- **The magnitude of the induced emf is proportional to the rate of change of flux linkage (Faraday's 2nd law of electromagnetic induction).**
- The direction of the induced emf can be reversed by either using the North pole of the magnet instead of the South or by withdrawing the magnet instead of inserting it.
- **The direction of the induced emf tends to oppose the change producing it (Lenz's Law).**

### Magnitude of the induced emf

In the qualitative demonstration, the size of the induced emf seems to be proportional to the "strength" of the magnet and to its speed. It is easy to visualise the magnetic field lines from the magnet (Figure 5.33) cutting through the wire turns of the coil. A bigger emf is induced when more field lines per second are cutting through the wire.

Thus:

$$\text{induced emf} \propto \text{rate of change of flux}$$

The size of the induced emf also seems to be proportional to the number of turns of wire on the coil. It is useful to define the **flux linkage** of the coil:

$$\text{flux linkage} = N\phi$$

where $\qquad\qquad$ $N$ = number of turns on the coil

$$\phi = \text{flux.}$$

So, for a coil with more than one turn:

$$\text{induced emf} \propto \text{rate of change of flux linkage}$$

or: $\qquad$ induced emf $= k \times$ rate of change of flux linkage

The size of the unit of flux (Wb) is consistent with the constant, $k$, having the value 1. Thus:

$$\text{induced emf} = \text{rate of change of flux linkage}$$

or: $\qquad$ $\text{induced emf} = \dfrac{\Delta(N\phi)}{t}$

### Direction of induced emf

Consider a magnet given a small velocity so that it moves horizontally into a coil (see Figure 5.50).

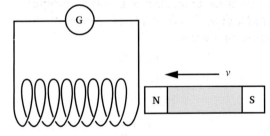

**Figure 5.50**

An emf induced in the coil causes a current to flow, which makes the coil into an electromagnet. The end of the coil next to the magnet must become either a N magnetic pole or a S magnetic pole.

Assume that a S pole is created near the magnet.
- The magnet will be attracted towards the coil and will accelerate.
- The rate of change of flux linkage will increase.
- The current dissipated in the circuit will increase as will the kinetic energy of the magnet.

From a very small input of kinetic energy in the first place, it seems that more kinetic energy is created at the same time as electrical energy is produced in the circuit. Clearly this breaches the law of conservation of energy.

The assumption that a S pole is created near the magnet must be incorrect: a N pole is created near to the magnet. This will oppose the movement which is causing the emf to be induced, i.e. the movement of the magnet into the coil. Hence the rather strange-sounding formulation of Lenz's Law that the direction of the induced emf tends to oppose the change producing it.

### Examples of calculations involving Faraday's Second Law

Consider a wire of length $l$ sliding down a conducting framework at right angles to a magnetic field of flux density $B$. The wire moves at a speed of $v$ and therefore travels $v$ m in 1 sec (see Figure 5.51).

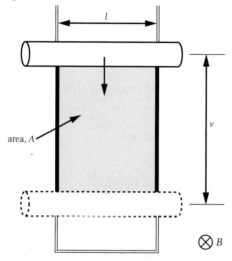

**Figure 5.51**

An emf, $E$, will be induced in the wire creating a current, $I$, in the frame.

$$E = \frac{\Delta(N\phi)}{t}$$

In this case $N = 1$ and $t = 1$,

So: $$E = \Delta\phi$$

The change in the flux linkage, $\Delta\phi$, is given by:

$$\Delta\phi = BA$$

$$= B.lv$$

so: $$E = Blv$$

This equation can be used to find, for example, the emf induced between the tips of an aeroplane's wings as it flies through the Earth's magnetic field.

### Applying Lenz's Law

The direction of the induced emf must tend to oppose the movement of the wire. A current flows in the wire because of the induced emf. The force on this current-carrying conductor in a magnetic field must be in an upward direction to oppose the

motion. Using Fleming's Left Hand Rule to find the current direction which would produce an upward force shows that the current must be going from left to right.

Alternatively, one could use Fleming's Right Hand Rule, which is specifically designed for this situation (see Figure 5.52).

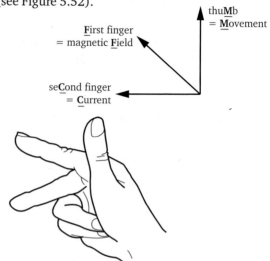

**Figure 5.52**

Using the rule on Figure 5.52 confirms that the current that is produced as a result of the induced emf is from left to right.

### Annoying way of remembering which of Fleming's rules to use

Whenever electricity and magnetism combine to produce a force or movement, the device concerned is behaving as a **motor**. Use the Left Hand Rule in these situations because **motor** cars drive on the **left**.

Whenever movement and magnetism combine to produce (or induce) electricity, the device concerned is a **generator**. Use the Right Hand Rule where **gene-right-ors** are concerned.

### Emfs induced in rotating wheels

Consider a wheel of radius $r$ spinning with angular velocity, $\omega$, in a magnetic field (see Figure 5.53).

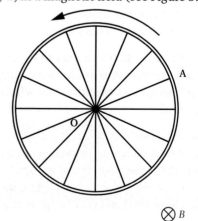

**Figure 5.53**

Since the spokes are cutting through the magnetic field of flux density $B$, an emf will be induced between the centre of the wheel and the rim.

$$E = \frac{\Delta(N\phi)}{t}$$

but

$$N = 1$$

so:

$$E = \frac{\Delta\phi}{t}$$

In 1 second, the wheel will have turned through an angle $\omega$ which is a fraction of a full rotation given by $\frac{\omega}{2\pi}$.

so:     area swept by OA in 1 sec $= \frac{\omega}{2\pi}.\pi r^2$

rate of change of flux $= \frac{\Delta\phi}{t}$

$$= \frac{BA}{t}$$

$$\frac{\Delta\phi}{t} = \frac{B\omega r^2}{2}$$

so:

$$E = \frac{B\omega r^2}{2}$$

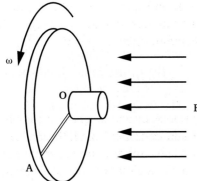

**Figure 5.54**

An identical calculation could be made for a solid wheel. By considering the element OA as similar to the spoke OA in the previous example, it can be shown that the emf induced between the centre and the rim of a solid wheel is given by:

$$E = \frac{B\omega r^2}{2}$$

## Induced emfs in Motors

The rotor of a motor turns because of the torque it experiences. The torque arises from the forces on the current carrying wires of the rotor which is set in a magnetic field.

However, when the rotor spins, it is also a conductor which is moving through a magnetic field. It will therefore have an emf induced in it. The direction of this induced emf (sometimes known as a back emf) will try to oppose the

movement of the rotor (Lenz's Law). Current is supplied to the rotor in opposition to this back emf. This supply of current against the back emf requires electrical energy to be used. It is this electrical energy that is converted into kinetic energy in the motor.

**Test your understanding**

1   Describe the energy changes that happen as the magnet is moved into the coil in Figure 5.50, recognising that a N pole is created at the right-hand side of the coil.

2   A 2.0 m long wire moves at right angles to a magnetic field of flux density 0.62 T at a speed of 0.43 m s$^{-1}$. Calculate the emf induced in the wire.

3   A 200 turn coil of cross-sectional area $4.2 \times 10^{-4}$ m$^2$ is situated in a changing magnetic field. The flux density changes according to the graph.

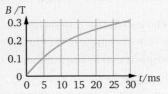

**Figure 5.55**

   (a)   Determine the flux linkage at time $t = 5$ ms.
   (b)   Estimate the rate of change of flux linkage at time $t = 5$ ms.
   (c)   State the emf induced in the coil.

4   A wheel of diameter 1.2 m spins at right angles to a magnetic field of flux density 0.68 mT. The wheel turns at a rate of 18 revolutions per second. Calculate the emf induced between the wheel's rim and its axle.

5   A plane flies through the Earth's magnetic field. Estimate the emf induced between the wing tips. State any assumptions you make and the values of any parameters that you use.

## Transformers

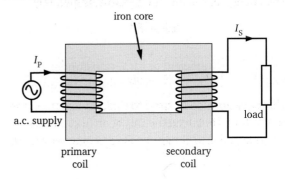

**Figure 5.56**

A transformer consists of a heavy iron core with two coils wrapped around it. The **primary** coil may be regarded as the input coil and is connected to an alternating current supply. The **secondary** coil is the coil from which the output is taken. The output voltage in the secondary coil is created as follows.

- The current in the primary coil creates a magnetic field in the core.
- Since iron is a good magnetic material, the flux density in the core is large.
- Since the primary current is changing constantly, the magnetic field in the core is changing constantly.
- The core ensures that the changing magnetic field is linked to the secondary coil.
- Since the magnetic field through the secondary coil is changing an emf is induced in the coil.

### Step up and step down transformers

The size of the emf induced in the secondary coil is equal to the rate of change of flux linkage. Flux linkage is proportional to the number of turns in the coil. Therefore, the output or secondary voltage $V_s$ from the transformer is proportional to the number of turns on the secondary coil. Transformers that have a larger output voltage than their input voltage are called step up transformers. For a step up transformer:

$$V_s > V_p$$

and:
$$N_s > N_p$$

where $V_s$ = secondary voltage

$V_p$ = primary voltage

$N_s$ = number of turns on secondary coil

$N_p$ = number of turns on the primary coil

Transformers which have a smaller output voltage than their input voltage are called step down transformers. For a step down transformer:

$$V_s < V_p$$

and:
$$N_s < N_p$$

For a perfectly efficient or ideal transformer, whether it is a step up or a step down transformer:

$$\frac{V_s}{V_p} = \frac{N_s}{N_p}$$

If a transformer is ideal, all power put in to the primary coil will be extracted from the secondary coil:

output power = input power
$$V_s I_s = V_p I_p$$
$$\frac{V_s}{V_p} = \frac{I_p}{I_s}$$

Notice that a transformer that steps up the voltage will have a lower secondary current than the primary current. Conversely if the secondary voltage is less than the primary voltage, then the secondary current is greater than the primary current,

if: $\qquad V_s > V_p \qquad$ then $\qquad I_s < I_p$

if: $\qquad V_s < V_p \qquad$ then $\qquad I_s > I_p$

### Transformer inefficiencies

### Copper losses or $I^2R$ losses

Whenever a current flows through a component with resistance, electrical energy is transformed into internal energy and the component is heated. In a transformer that is supplying a secondary current, both coils will be heated. Copper losses can be minimised by using a good conducting material (such as copper) for the coils and by using wire of large cross-sectional area to minimise the resistance of the coils (see resistivity: Module 1).

### Flux losses

Not all of the flux created by the primary coil links with the secondary coil, resulting in a value for $V_s$ which is less than anticipated. This is minimised by careful design of the core in which the secondary coil is wound on top of the primary coil (see Figure 5.57).

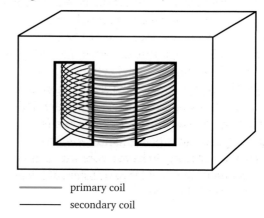

——————— primary coil
——————— secondary coil

***Figure 5.57***

### Eddy currents

An emf is induced in the secondary coil because it is a conductor in the presence of a changing magnetic field. The iron core is also a conductor in the presence of a changing magnetic field so an emf is induced in the core itself. Currents circulate or eddy around the iron core as a result of these emfs. As with any other situation in which there is a current in a material which has resistance, electrical energy is transferred into internal energy. Energy is wasted heating the core.

Eddy current losses can be minimised by making sure that the eddy currents are as small as possible. Since the power loss is equal to $I^2R$, reducing the current reduces the power loss. In transformers, the eddy currents are made smaller by dividing the iron core into sheets insulated from each other by lacquer or varnish. The strips only accommodate small eddy currents, whereas a chunky core would sustain large eddy currents. This is called laminating the core (see Figure 5.58).

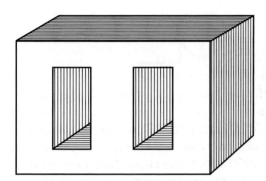

**Figure 5.58**

### Hysteresis losses

A hard magnetic material (such as steel) can be magnetised and retain its magnetism. A soft magnetic material (such as iron) can be magnetised, but loses its magnetism very easily. This is what makes iron suitable for use in the core of a transformer, where it is necessary to reverse the direction of the magnetic field every time the primary current changes direction. However, iron is not a perfectly soft magnetic material: once magnetised in one direction it resists being magnetised in the opposite direction. Reversing the direction of the magnetic field in the core takes energy. The energy used by the primary coil to reverse the magnetic field heats up the core. This so-called hysteresis loss can be minimised by keeping the flux density low. With a low flux density, it is necessary to have a core with a large cross-sectional area in order to transfer sufficient flux to the secondary coil.

### Calculation involving transformer efficiency

Transformer inefficiencies result in reductions to the output voltage and the output current. It is difficult to predict the ways in which both of these parameters will be affected. For example, copper losses in the secondary coil will be observed as an internal resistance in the output side of the transformer (see Module 1). If a large current is being drawn from the secondary coil of the transformer, this will result in a large "lost volts" effect: the output voltage will be reduced below the value that the turns ratio would predict.

As with any other efficiency situation:

$$\text{efficiency} = \frac{\text{useful power output}}{\text{total power input}}$$

$$\text{efficiency} = \frac{V_s I_s}{V_p I_p}$$

### Use of transformers for transmission of electricity

Transmission of electricity through the grid involves transformers because:
- transmission is executed at high voltage
- if the voltage is high, large powers can be delivered using relatively low currents
- if the current is low, the power losses in the cables are low
- high voltages are inappropriate for industrial or domestic electricity consumption
- transformers can raise the voltage for transmission and reduce it for consumption.

Calculations of power losses are covered in Module 1. Examples of calculations of power losses in the national grid are covered in the AS text.

### Test your understanding

1  Fill in the missing details in the following table. All apply to perfectly efficient transformers.

**Table 5.8**

|   | $N_P$ | $N_S$ | $V_P$ V | $V_S$ V |
|---|---|---|---|---|
| a | 2400 | 600 | 12 | |
| b | 6000 | 300 | | 12 |
| c | 4500 | | 230 | 4.0 |
| d | | 2000 | 12 | 230 |

2  For the perfectly efficient transformers in (**1**):
  (a) calculate $I_p$, the input power and the output power if $I_S$ is 2.0 A
  (b) calculate $I_S$, the input power and the output power if $I_p$ is 0.36 A
  (c) calculate $I_S$, $I_p$ and the output power if the input power is 8.6 W
  (d) calculate $I_S$, $I_p$ and the input power if the output power is 60 W.

3  A step down transformer draws a current of 0.12 A from the 230 V mains supply. The transformer's turns ratio is 1/20. Inefficiency reduces the secondary voltage to 95% of that which is expected.
  (a) Calculate the secondary voltage.

  The secondary coil has a resistance of 0.50 Ω.
  The transformer's load has a resistance of 6.5 Ω.
  Calculate:
  (b) the secondary current
  (c) the power output
  (d) the efficiency.

## Eddy Currents in Other Devices

As well as in transformer cores, eddy currents arise in motors. The rotor is wound on a metallic former that spins in the magnetic field produced by the magnets. Eddy currents in the former will heat the material in which they occur leading to inefficiency in the motor. Eddy currents can be reduced by laminating the former.

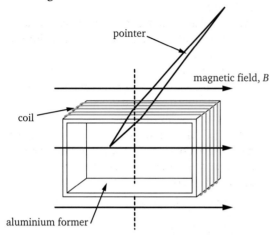

**Figure 5.59**

Older electrical meters use pointers attached to coils of wire that twist in a magnetic field in a similar way to the rotor of a motor. The coil is wound on to an aluminium former which is also subject to eddy currents (see Figure 5.59). The eddy currents are a positive advantage in these meters as they stop the coil swinging around as the pointer settles on to its final reading. Kinetic energy from the moving former and coil is dissipated as internal energy. The movement of the coil is said to be damped.

## Simple Alternator

Figure 5.60 is a schematic diagram of a simple alternator. Notice how similar it is to a simple motor. In fact, motors can behave as generators and vice versa. The alternator is a generator that produces alternating current, whereas the motor shown in Figure 5.41 (page 105) is a direct current device.

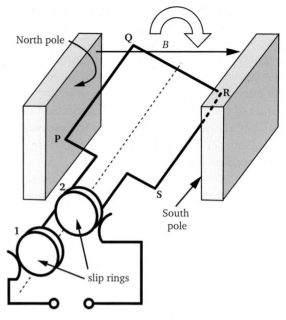

**Figure 5.60**

### Why the alternator produces a.c.

As the coil, **PQRS**, rotates in the magnetic field, an emf is induced in the coil. Fleming's Right Hand Rule can be used to determine the direction of the current from the output of the alternator.

Consider **PQ** moving upwards as shown in Figure 5.60. When a load is connected across the output terminals, the current will be from **P** to **Q**. The current will be entering the rotor through slip ring **1** and leaving it through slip ring **2**. The right-hand terminal will be positive.

When the rotor has turned far enough for **PQ** to be moving downwards, the current direction will be reversed. The left-hand terminal will now be positive. Each complete turn of the rotor produces one cycle of alternating current (a.c.). The frequency of the a.c. is the same as the frequency of rotation of the rotor.

### The variation of emf with time

As with any induction phenomena, the magnitude of the induced emf is equal to the rate of change of flux linkage. There are two ways of assessing how the magnitude of the output voltage will vary with time.

Firstly, consider the movement of conductor **PQ**. When the rotor is horizontal, **PQ** is cutting directly through the magnetic field at right angles to it. At this point, the induced emf in both **PQ** and **RS** will be a maximum. When the rotor is vertical, although **PQ** and **RS** will be moving at the same speed, they will be moving parallel to the magnetic field, not cutting through it. At this point, the induced emfs and the output voltage will be zero.

Secondly, consider the rotor with its area $A$. $A$ is equal to $lb$. If the coil has only one turn on it, the maximum flux linkage the coil will have with the field occurs when the coil is vertical and presents its whole area to the field:

$$\text{flux linkage} = BA$$

The flux linkage will vary according to the cosine wave shown in the graph in Figure 5.61. $\theta$ is the angle between the normal to the coil and the magnetic field $B$. The diagram also shows the rotor positions and the variation of output voltage.

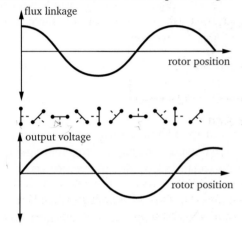

**Figure 5.61**

The rate of change of flux linkage is the gradient of the top graph. Although the flux linkage is maximum when the coil is vertical, the rate of change of flux linkage is zero and so the output voltage is zero. When the flux linkage itself is zero, the gradient of the top graph is a maximum, showing the greatest rate of change of flux linkage and so the maximum output voltage.

## Factors affecting the peak output voltage

Since the voltage is equal to the rate of change of flux linkage, the peak output voltage is proportional to:

- the flux density of the magnetic field;
- the area of the rotor;
- the rate of rotation of the rotor;
- the number of turns of wire on the rotor.

### Test your understanding

1 Explain in terms of flux linkage why the factors stated above affect the peak output voltage of a simple alternator.

2 An alternator such as the one in Figure 5.60 has a rotor of length 5.0 cm and breadth 3.0 cm. The magnetic field is of flux density 0.80 T. There are 20 turns on the rotor, which turns at a rate of 12 revolutions per second.
   (a) Find the speed of the conductor **PQ**.
   (b) Calculate the emf induced in one turn of **PQ** when the rotor is horizontal.
   (c) State the emf induced in the rotor at this moment.

   (d) Sketch a graph showing the variation of emf with rotor position.
   (e) State and explain how the output would change if the speed of rotation were reduced to 4.0 revolutions per second.

3 A transformer is required to produce an r.m.s. output of $2.0 \times 10^3$ V when it is connected to the 230 V r.m.s. mains supply. The primary coil has 800 turns.
   (a) Calculate the number of turns required on the secondary coil, assuming the transformer is ideal.
   (b) The transformer suffers from *eddy current* losses. Explain how *eddy currents* arise. State the feature of transformers designed to minimise eddy currents.

4 A vehicle braking device consists of an aluminium disc attached to the road wheel. Electromagnets on either side of the disc can cause an emf to be induced in the disc.

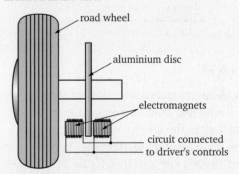

**Figure 5.62**

   (a) The disc, of radius 0.23 m, rotates at an angular speed of 83 rad s$^{-1}$. The electromagnets provide a uniform field of flux density 1.6 T. Assuming that the emf covers the whole area of the disc, calculate the emf induced between the rim and the centre of the axle.
   (b) (i) Explain how the induction of an emf causes the vehicle to slow down.
       (ii) Explain why the braking device will become less effective at low speed.
   (c) Draw a circuit diagram to show how the braking force could be varied by the driver.

5 Solar storms consist of protons and helium nuclei emitted from the sun. Protons from a storm hit the Earth's atmosphere at a speed of $1.2 \times 10^6$ m s$^{-1}$. The particles pass through a region where the Earth's magnetic field is vertically downwards and has a flux density of $5.8 \times 10^{-5}$ T.
   (a) Calculate the radius of the proton's path. (Proton mass is $1.7 \times 10^{-27}$ kg and charge is $1.6 \times 10^{-19}$ C).
   (b) Draw a diagram to show the path of the proton. On the same diagram, show the path of a helium nucleus that has the same initial velocity as the proton. State the radius of the path of the helium nucleus. (Helium nucleus mass is $6.8 \times 10^{-27}$ kg and charge is $3.2 \times 10^{-19}$ C.)

# Nuclear Energy

## The Strong Nuclear Force

A nucleus consists of protons and neutrons packed together into a very small space. Protons and neutrons are separated by distances of about 1 fm (1 femtometre).

$$1 \text{ fm} = 10^{-15} \text{ m}$$

Electrostatic repulsion will be experienced between adjacent protons. The force of repulsion will be given by:

$$F = \frac{Q_1 Q_2}{4\pi\varepsilon_0 r^2}$$

$$Q_1 = Q_2 = 1.6 \times 10^{-19} \text{ C}$$

$$\varepsilon_0 = 8.9 \times 10^{-12} \text{ Fm}^{-1}$$

$$r = 1.0 \times 10^{-15} \text{ m}$$

$$F = \frac{1.6 \times 10^{-19} \times 1.6 \times 10^{-19}}{4\pi \times 8.9 \times 10^{-12} \times (1.0 \times 10^{-15})^2}$$

$$F = 230 \text{ N}$$

Considering that a proton has a mass of only $1.7 \times 10^{-27}$ kg, this is a very large force to be exerted on a small particle.

Since both protons and neutrons have mass, they will experience mutual gravitational attraction. The force of attraction is given by:

$$F = \frac{Gm_1 m_2}{r^2}$$

$$G = 6.7 \times 10^{-11} \text{ N m}^2 \text{ kg}^{-2}$$

$$m_1 = m_2 = 1.7 \times 10^{-27} \text{ kg}$$

$$r = 1.0 \times 10^{-15} \text{ m}$$

$$F = \frac{6.7 \times 10^{-11} \times 1.7 \times 10^{-27} \times 1.7 \times 10^{-27}}{(1.0 \times 10^{-15})^2}$$

$$F = 1.9 \times 10^{-34} \text{ N}$$

This is 36 orders of magnitude smaller than the electrostatic repulsion that exists between the protons. Clearly, gravity is insufficient to hold the nucleus together against the electrostatic repulsion of the protons. There must be another force that does this job.

The new force is called the strong nuclear force or, sometimes, the strong nuclear interaction. The properties of the strong nuclear force were deduced in the 1930s.

- At distances of about 1 fm it must be strongly attractive to hold the nucleus together against the electrostatic repulsion of the protons (see Figure 5.63).
- It is not dependent on charge as it affects both protons and neutrons in the same way. In fact, it affects all hadrons (baryons and mesons) and quarks. It is responsible for holding quarks together in mesons and baryons (Module 2).
- At distances of less than about 1 fm it must be a repulsive force (see Figure 5.63). If it stayed attractive and got stronger with decreasing distance, nuclei would collapse into immeasurably small volumes.
- At distances of greater than a few fm it must disappear, otherwise it would be detectable outside the nucleus. As it is, in Geiger and Marsden's alpha particle scattering experiment (Module 2), the deflection of the alpha particles is explained by the existence of only electrostatic repulsion between the alpha particles and the gold nuclei.
- Neighbouring nucleons seem to 'shroud' the force so that the strong nuclear force is not experienced by other nucleons that are slightly further away.

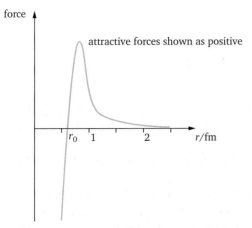

**Figure 5.63**

Figure 5.63 shows a graph of the variation of the strong nuclear force with separation, $r$, of nucleons.

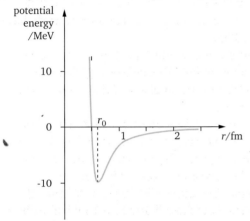

**Figure 5.64**

Considering only the strong nuclear force, as two nucleons are brought together from infinity, their potential energy will decrease. As the strong nuclear force is attractive, work would need to be done to separate the nucleons. See page 98 for a similar argument relating to potential energy in gravitational fields. If the nucleons are pushed closer together than about 1 fm, their potential energy becomes positive since work would have to be done against the strong nuclear force, which at these distances has become repulsive. Figure 5.64 shows the variation of potential energy against separation for the strong nuclear force. If the strong nuclear force were the only force affecting the nucleons, they would adopt a separation of $r_0$ where their potential energy is a minimum. Notice also that, at $r_0$, the force between the nucleons is zero.

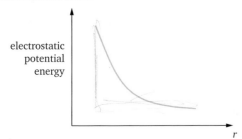

**Figure 5.65**

In practice, the nucleons are affected by the strong nuclear force and by electrostatic repulsion. Figure 5.65 shows a qualitative potential energy curve for the potential energy in the electrostatic field between a proton and a nucleus. Putting the two potential energy curves together and showing the curves for each side of the nucleus gives Figure 5.66.

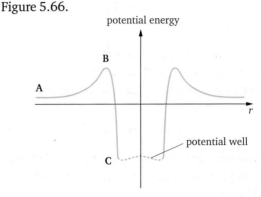

**Figure 5.66**

Notice:
- at **A** the electrostatic force dominates making the potential energy positive
- at **B** the strong nuclear force is beginning to take over, attracting the nucleons together so that the potential energy gets less
- at **C** the potential energy is a minimum so this will be where the nucleons are at their equilibrium separation. (In this diagram, the curve has not been defined for smaller separations).

When a nucleus is constructed, the nucleons have a negative potential energy. This makes the nucleus stable because to take it to bits would require energy to be supplied. The nucleons would need to be lifted out of the so-called *potential well* that they are in. The energy that would need to be supplied to dismantle the nucleus is called the **binding energy**.

> **Test your understanding**
>
> 1  Look at Figures 5.63 and 5.64. The relationships between force and potential energy for the strong nuclear force are like the relationships between force and energy for electrostatic forces (Figure 5.19).
>    (a)  Why does the force become negative at distances less than $r_0$?
>    (b)  Explain why the force is zero at $r_0$.
>    (c)  Explain why the potential energy is a minimum at $r_0$.
>    (d)  Use data from Figure 5.64 to estimate the **maximum** force between two protons.

## Mass Defect

A $^{235}_{92}$U nucleus is made up of 92 protons and 143 neutrons. Consider the total mass of the constituents compared with the mass of the nucleus:

$$\text{mass of a proton} = 1.673 \times 10^{-27} \text{ kg}$$
$$\text{mass of a neutron} = 1.675 \times 10^{-27} \text{ kg}$$

so: mass of constituents $= (92 \times 1.673 \times 10^{-27})$
$$+ (143 \times 1.675 \times 10^{-27})$$
$$= 3.9344 \times 10^{-25} \text{ kg}$$

$$\text{mass of } ^{235}_{92}\text{U nucleus} = 3.9046 \times 10^{-25} \text{ kg}$$

It seems that the nucleus has less mass than the bits it is made of! The difference is usually called a **mass defect** and, in this case, is $3.0 \times 10^{-27}$ kg.

We noted in the previous section that the nucleons had lost energy in order to be assembled into the nucleus. In fact, matter and energy are interchangeable and the reduction in mass is the observed effect of the particle's loss of energy. The mass defect is the equivalent of the binding energy of the nucleus.

The mathematical connection between matter and energy is given in Einstein's equation:

$$E = mc^2$$

where  $E$ = energy released

$m$ = the mass that has been lost

$c$ = the speed of electromagnetic radiation

The total binding energy of the uranium nucleus can be calculated:

$$E = mc^2$$

where  $m = 3.0 \times 10^{-27}$ kg
$c = 3.0 \times 10^8$ m s$^{-1}$

so:   $E = 2.7 \times 10^{-10}$ J

## Examples of Interchangeability of Matter and Energy

Matter and energy are mutually interchangeable. If a body gains energy, its mass increases. The change is usually far too small to be noticed. The increase in mass of a kettle full of water as it is heated to boiling temperature would not be noticed. However, when dealing with interactions between particles of very small mass, changes of the size calculated above are easily noticeable.

If a particle of matter meets an appropriate particle of antimatter they will annihilate each other (see Module 2). The matter will disappear and be replaced by energy in the form of electromagnetic radiation.

When an electron and a positron, each of mass $9.1 \times 10^{-31}$ kg meet and annihilate each other, two gamma rays are produced. The combined energy of the gamma rays is:

$$E = mc^2$$

where   $m = 2 \times 9.1 \times 10^{-31}$ kg
$c = 3.0 \times 10^8$ m s$^{-1}$

so:     $E = 2 \times 9.1 \times 10^{-31} \times (3.0 \times 10^8)^2$

$$E = 1.6 \times 10^{-13} \text{ J}$$

This interchangeability of mass and energy can be seen in various contexts.

- Particle annihilation releasing energy (see above).
- A particle decaying spontaneously into other particles, releasing energy as the decay occurs, e.g. a neutron decaying into a proton and an electron.
- Radioactive decay, e.g. the emission of an alpha particle from a nucleus. In this case, the energy appears as the kinetic energy of alpha particles and of the remaining nucleus as it recoils.
- The fission of the nucleus into smaller nuclei and other particles. Once again, most of the energy produced is in the form of the kinetic energy of the particles.
- Fusion of two light nuclei into one heavier one.

In all of these examples, the mass of the products must be less than the mass of the original particles. The change in mass is equivalent to the amount of energy released.

Other examples of matter/energy interchangeability include:
- the spontaneous creation of a particle/antiparticle pair from a photon of electromagnetic radiation
- the creation of very massive particles from the collision of two light particles that have large KE before the collision, e.g. the creation of W particles in a particle accelerator.

In each of these examples, the products have greater mass than was present before the event. The extra mass is equivalent to the energy which has gone into the production of the particles.

## Units Used in Nuclear Physics

Since the masses involved in Nuclear Physics are very small, it is often convenient to use the atomic mass unit (u) instead of the kg. One atomic mass unit is $\frac{1}{12}$ of the mass of a carbon-12 ($^{12}_{6}C$) atom.

$$1\,u = 1.66 \times 10^{-27}\,kg$$

Similarly, the values of energy used are very small by everyday standards. So electron volts (eV) or mega-electron volts (MeV) are often used. An electron volt is the energy acquired by an electron when it is accelerated through a potential difference of 1 V.

Since, for charged particles moving through electric fields:

$$Energy = Vq$$

$$1\,eV = 1\,V \times 1.6 \times 10^{-19}\,C$$

$$1\,eV = 1.6 \times 10^{-19}\,J$$

$$1\,MeV = 1.6 \times 10^{-13}\,J$$

If, during an event, a mass of 1 u is lost, the energy produced is given by:

$$E = mc^2$$

where
$$m = 1\,u = 1.66 \times 10^{-27}\,kg$$
$$c = 3.0 \times 10^8\,ms^{-1}$$
$$E = 1.49 \times 10^{-10}\,J$$
$$= \frac{1.49 \times 10^{-10}}{1.6 \times 10^{-19}}$$
$$= 9.3\ 10^8\,eV$$
$$= 930\,MeV$$

## The Binding Energy Curve

Not all nuclei have the same binding energy. The binding energy for a uranium nucleus was calculated at the beginning of this section and found to be $2.7 \times 10^{-10}$ J. Converting this to MeV gives a binding energy of 1690 MeV. Since there are 235 nucleons in a $^{235}_{92}U$ nucleus:

$$binding\ energy\ per\ nucleon = \frac{1690}{235}$$
$$= 7.2\,MeV$$

If similar calculations are performed for many nuclei and the binding energy per nucleon is plotted against the nucleon number, the result is seen in Figure 5.67.

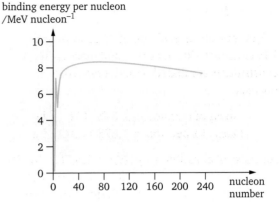

**Figure 5.67**

## Energy from Fission

If a heavy nucleus can be made to split, it will produce two lighter nuclei. Each of these new nuclei will have a greater binding energy per nucleon (see Figure 5.68).

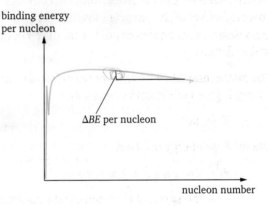

**Figure 5.68**

Remember that the binding energy is equivalent to the mass defect of the nucleus. The fact that the new nuclei have greater binding energies than the original nucleus means that their total mass is less than that of the original nucleus. The change in mass is equivalent to the energy release that accompanies the fission.

1 Use a spreadsheet to work out binding energies per nucleon for 20 nuclei over the full range of masses. Include $^4_2$He, $^2_1$H and $^6_3$Li in your calculations. Use your answers to plot a graph like Figure 5.67. You will need to use a data book to find the necessary masses. Remember that data books give atomic masses – to find nuclear masses you will need to subtract the mass of an appropriate number of electrons.

2 Mark on to your graph and explain the significance of the positions of the following nuclei: '

   (a) $^2_1$H

   (b) $^4_2$He

   (c) $^{54}_{26}$Fe

   (d) $^{235}_{92}$U.

## The Fission Process

A few heavy isotopes are fissile. This means that they are capable of splitting into smaller nuclei and releasing energy. The most commonly used fissile isotope is uranium 235 ($^{235}_{92}$U). Uranium will occasionally undergo spontaneous fission, but a nucleus can be induced to fission by making it absorb a neutron. When $^{235}_{92}$U absorbs a neutron, the nucleon number increases by one but the proton number is unchanged.

$$^{235}_{92}U + ^1_0n \rightarrow ^{236}_{92}U$$

$^{236}_{92}$U is unstable and will split into two parts or fission products. Each fission product is roughly half of the $^{236}_{92}$U nucleus. Fission products are always radioactive, usually β-emitters. As the nucleus splits, a few neutrons are released. On average 2.4 neutrons are produced per fission.

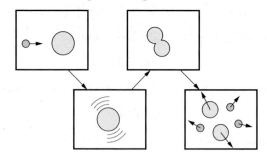

**Figure 5.69**

Since the sum of the products' masses is less than the mass of the $^{236}_{92}$U, energy is released when the fission happens. This energy is released in a variety of forms:

- KE of the fission products
- KE of the neutrons

- energy released subsequently when fission products undergo radioactive decay
- energy carried away by neutrinos (in practice this is lost as the neutrinos will rarely interact with other matter)
- γ rays released at the time of fission.

It is the release of energy that makes the nuclear fission process of commercial interest. It is the fact that each fission is accompanied by the emission of a few more neutrons that makes it practicable to exploit the fission process. Since the absorption of a neutron is required to instigate fission, the production of neutrons during fission allows the following generation of fissions to occur (see Figure 5.70).

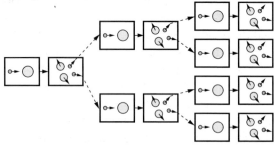

**Figure 5.70**

As each fission is "caused" by the previous fission, this is called a chain reaction.

## Nuclear Fission Reactors

Although, for examination purposes, you do not need to know the details of any particular type of nuclear reactor, it is necessary to know the function of the *moderator*, the *coolant* and the *control rods*. The reactor described here is called an Advanced Gas Reactor (AGR) and is in current use in Britain.

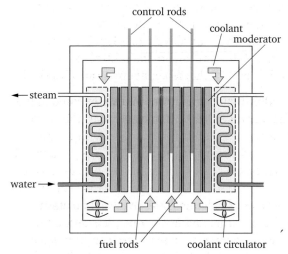

**Figure 5.71**

**Fuel rods:** These are made from uranium oxide. Only 0.7% of naturally occurring uranium is the main fissile isotope, $^{235}_{92}$U. This is 'enriched' so that 5% of the uranium in the fuel rod is $^{235}_{92}$U.

Fissions occur in the fuel rod. Fission products collide with other atoms in the rod increasing their internal energy, i.e. the temperature of the rods increases as a result of the fissions.

**Coolant:** Carbon dioxide gas is pumped rapidly around the reactor by the gas coolant circulators. It is heated by the fuel rods as it passes over them. The coolant then passes through heat exchangers. Water passing through pipes in the heat exchangers, is boiled and the steam is super-heated by the energy from the carbon dioxide gas. The super-heated steam is used to drive turbines linked to generators for electricity production.

**Control rods:** The chain reaction shown in Figure 5.70 is growing or supercritical. If all of the reaction chains in a nuclear reactor were like this, the reaction would get out of control very quickly! In a steady, stable situation each fission that happens would lead to only one fission in the next generation of the reaction. On average, 2.4 neutrons are produced by each fission. On average exactly one neutron must go on to cause a fission in the next generation. A steady chain-reaction of this type is called **critical**. To keep the reaction critical, an average of 1.4 neutrons must be lost from each fission. There are many ways in which neutrons can be lost – they can escape entirely from the reactor or can be absorbed in a non-fission event by some material in the reactor.

A reactor that lost more than 1.4 neutrons per fission could not sustain a reaction. Reactors are carefully designed to lose as few neutrons as possible. To keep the reaction exactly critical, control rods made from a neutron-absorbing material, such as boron, are used. When it is necessary to slow the reaction down, the control rods are lowered further into the reactor, so that more neutrons are absorbed. When it is necessary to speed the reaction up, the control rods are withdrawn, so that fewer neutrons are absorbed.

**Moderator:** The neutrons that are released in the fission process are very fast neutrons: they have kinetic energies of the order of MeV. Such fast neutrons are unlikely to be absorbed by uranium in a fission event so the neutrons need to be slowed down. Neutrons will slow down if they collide with particles of a similar mass. In a head on elastic collision with a hydrogen atom, the neutron will lose all its energy in one go (see conservation of momentum calculations in Module 4). It would not be a good idea to surround the fuel in a reactor with hydrogen gas, but if the fuel rods are surrounded with water, there will be plenty of hydrogen atoms with which the neutrons can collide in order to slow

down. Other materials that are used for this purpose include deuterium (heavy hydrogen) in the form of heavy water, and carbon in the form of graphite (as used in AGR reactors). Materials used for the purpose of slowing down the neutrons are known as moderators. In the interests of neutron economy, moderators should not be good absorbers of neutrons.

---

**Test your understanding**

1   Compare the effectiveness of heavy water, water and graphite as moderators in a nuclear reactor. Consider their effectiveness in slowing down neutrons and their neutron absorbencies. The following question is extension work and involves tricky mathematics.

2   Use conservation of momentum and conservation of energy to work out how many head-on, elastic collisions it would take for a neutron, travelling at $2.0 \times 10^7$ m s$^{-1}$, to slow to $1.0 \times 10^3$ m s$^{-1}$ by having head-on collisions with:
    (a)  deuterium atoms;
    (b)  carbon atoms.
    (Mass of carbon atom = 12 u, mass of deuterium atom = 2 u.)

---

## Calculating the Energy Released from Fission

The energy released from a fission is equivalent to the mass change that takes place. For the purpose of calculation, the masses of complete atoms may be used, as the electron population is unchanged during the fission: relative atomic masses are more readily available than nuclear masses.

One possible fission reaction is:

$$^{235}_{92}U + ^{1}_{0}n \rightarrow ^{236}_{92}U \rightarrow ^{144}_{56}Ba + ^{90}_{36}Kr + 2^{1}_{0}n$$

mass of $^{235}_{92}U + ^{1}_{0}n = 235.124 + 1.009$

$$= 236.133 \text{ u}$$

mass of products $= 143.923 + 89.920 + 2(1.009)$

$$= 235.861 \text{ u}$$

change in mass $= 236.133 - 235.861$

$$= 0.272 \text{ u}$$

since a mass loss of 1u produces 930 MeV,

energy released in fission $= 0.272 \times 930 \text{ MeV}$

$$= 253 \text{ MeV}$$

$$= 4.0 \times 10^{-11} \text{ J}$$

Note that some of the data used is to six significant figures. This is unusual, but has to be done in these calculations as the final changes of mass are very small: the use of three significant figure data would mask the change.

## Fusion Reactions

Figure 5.72 shows that, if two light nuclei can be made to join together, the resulting nucleus will have a greater binding energy per nucleon than the original nuclei. When two nuclei join together, the process is known as fusion.

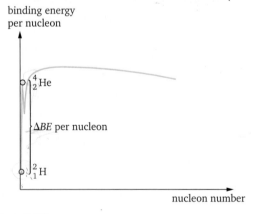

**Figure 5.72**

Since there is an increase in binding energy per nucleon, it should be the case that the new nucleus is lighter than the two nuclei that joined. The fusion shown in Figure 5.72 can be written:

$$^{2}_{1}H + ^{2}_{1}H \rightarrow ^{4}_{2}He + Energy$$

The energy released by the fusion can be determined by finding the change in mass during the reaction:

$$\text{mass of } ^{4}_{2}He = 4.0026 \text{ u}$$

$$\text{mass of } ^{2}_{1}H = 2.0141 \text{ u}$$

$$\text{change in mass} = (2 \times 2.0141) - 4.0026$$

$$= 0.0256 \text{ u}$$

Since a change in mass of 1 u releases 930 MeV:

$$\text{Energy} = 24 \text{ MeV}$$

Had the mass of the products been greater than the mass of the fusing nuclei, the reaction would not have been viable.

## Getting Fusion to Happen

For nuclei to fuse, they need to come into contact. Firstly, the atoms must be ionised. If a gas is heated enough, the electrons will break from the atoms creating a plasma. Individual nuclei in the plasma will experience electrostatic repulsion. To overcome this, the nuclei need to be travelling quickly.

Consider one hydrogen nucleus to be stationary. Another nucleus is aimed at the first. As this nucleus approaches it will lose kinetic energy and gain potential energy. If the moving nucleus has insufficient initial kinetic energy, it will stop before it gets close enough to the other nucleus to fuse. The nuclei need to come within about 1 fm of each other to fuse. The initial kinetic energy of the moving nucleus must be greater than or equal to the potential energy stored between the two nuclei at this separation. In other words the plasma needs to be at a very high temperature for the particles to be moving fast enough to come into contact. The temperatures needed are of the order of $10^8$ K.

## Feasibility of Radioactive Decays

You can check whether any particular decay is possible by comparing the sum of the masses of the products with the mass of the original atom. Consider the following reaction:

$$^{226}_{88}\text{Ra} \rightarrow {}^{222}_{86}\text{Rn} + {}^{4}_{2}\text{He}$$

mass of $^{226}_{88}\text{Ra}$ = 226.0254 u

mass of $^{222}_{86}\text{Rn}$ = 222.0175 u

mass of $^{4}_{2}\text{He}$ = 4.0026 u

mass at start of reaction = 226.0254 u

mass at end of reaction = 222.0175 + 4.0026

$$= 226.0201 \text{ u}$$

The mass of the products is less than the mass of the original radium atom so the reaction can, and will, proceed. The energy, $E$, released will be:

$$E = (226.0254 - 226.0201)\ 930$$
$$= 4.9 \text{ MeV}$$

### Test your understanding

1 (a) Uranium-238 decays by alpha emission to thorium-234. The table shows the masses of the nuclei of uranium-238 ($^{238}_{92}\text{U}$), thorium-234 and an alpha particle (helium-4).

**Table 5.9**

| Element | Nuclear mass/u |
|---|---|
| Uranium-238 | 238.0002 |
| Thorium-234 | 233.9941 |
| Helium-4, alpha particle | 4.0015 |

   (i) How many neutrons are there in a uranium-238 nucleus?

   (ii) How many protons are there in a nucleus of thorium?

  (b) (i) Determine the mass change in kg when a nucleus of uranium-238 decays by alpha emission to thorium-234.

   (ii) Determine the increase in kinetic energy of the system when a nucleus of uranium-238 decays by alpha emission to thorium-234.

2 (a) Calculate the mass change in kg when a tritium ($^{3}_{1}\text{H}$) nucleus is formed from its constituent parts.

Mass of
tritium nucleus = 3.016050 u
Mass of proton = 1.007277 u
Mass of neutron = 1.008665 u
Atomic mass unit, u = 1.660566 × $10^{-27}$ kg
Speed of light = 3.0 × $10^8$ m s$^{-1}$

  (b) Calculate the binding energy, in J, of the tritium nucleus.

# Particle Accelerators and Detectors

## Particles

Module 2 referred to the particles that appear to be the 'bricks' that make up atoms and their nuclei. In this section, we shall have a look at some of the experimental techniques that have been used to provide evidence about these particles and their properties.

## The Cloud Chamber

One of the most useful tools for investigating particles is the cloud chamber. All atomic and nuclear particles are very small – of the order of $10^{-13}$ m or much less; the wavelength of violet light is $4 \times 10^{-7}$m, so there is no possible chance of seeing them.

The cloud chamber makes use of the fact that if the pressure of a saturated vapour is suddenly reduced, some of the vapour condenses on any available object. In clouds in the atmosphere, water vapour often condenses on dust particles. In the cloud chamber, ionised atoms provide the opportunity for vapour to condense.

When electrons, protons or other charged atomic particles move through a gas, they ionise the atoms in their path. In the right circumstances, vapour will condense on these ions making the path visible.

You will not be examined on types of cloud chamber. Details of this one are included because the way in which it encourages the formation of drops uses principles from Module 4. Figure 5.73 shows the main features of the cloud chamber.

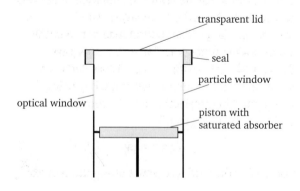

*Figure 5.73*

- The transparent lid enables viewing or photography.
- The optical window admits light.
- The particle window represents any arrangement for getting the particles into the chamber.
- The absorber is soaked with a volatile fluid, such as alcohol, to provide the saturated vapour.
- The piston is pulled down sharply. Expansion of the gas in the chamber is adiabatic and so both the pressure and temperature in the chamber fall sharply. Some of the vapour condenses on any ion tracks left by particles travelling through the chamber.

### Identifying particles

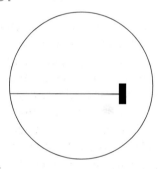

*Figure 5.74*

Figure 5.74 shows a track from a beta particle or electron track. The characteristic track of a very fast beta particle is long, straight and fine. It is fine because it does not hit many atoms along its path – it can therefore go a long way before it loses its energy. Slower beta particles produce tortuously curved tracks as they are more easily deviated in collisions.

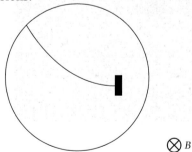

*Figure 5.75*

Figure 5.75 shows the track when there is a magnetic field going down into the plane of the diagram at the time of formation of the track.

The radius of curvature, $r$, of the track is given by:

$$r = \frac{mv}{qB}$$

or:

$$r = \frac{\text{particle momentum}}{qB}$$

where $q$, $m$ and $v$, are the charge, mass and velocity of the particle respectively, and $B$ is the flux density of the applied magnetic field (see page 107). Clearly, if the track can be identified, $q$ and $m$ will be known, and the velocity of the particle can be deduced from the radius of curvature of the track.

## Particle interactions

The usefulness of the cloud chamber is much wider than that suggested by the simple example quoted above. Many of the particles involved in nuclear physics have been discovered, and their properties elucidated, from tracks obtained in cloud chambers or in a similar device: the bubble chamber.

## The Bubble Chamber

A bubble chamber is a tank containing a liquid such as hydrogen maintained at or above its boiling point (made possible by high purity and cleanliness). 'Boiling' occurs along the tracks of charged particles as small bubbles of gas are formed at the ions that have been made along it. As liquid densities are higher than gaseous densities – more atoms per unit length of track – it can be a more sensitive device than the cloud chamber.

Figure 5.76 is a photograph of a bubble chamber in which an antiproton has collided with a proton. Pions and muons have been produced.

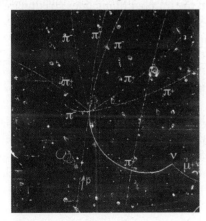

**Figure 5.76**

The importance of this example is that it illustrates another analytical tool available to the physicist – the application of the principle of the conservation of momentum. By measuring the radius of the circular paths of the particles, the charge and/or momentum of the particles can be determined.

Figure 5.77 shows a fine beam tube. These and other types of tubes, in which beams of electrons were subjected to magnetic and/or electric fields, enabled early measurements of the charge to mass ratio ($e/m$) of the electron to be made.

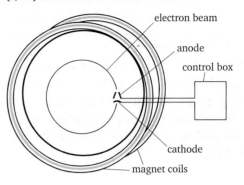

**Figure 5.77**

A beam of electrons is produced in the evacuated tube. The beam glows, making it possible to make measurements of it. The coils on the side of the tube produce a magnetic field, $B$. $B$ can be found by measuring the coils and the current in them. The diameter of the visible beam of electrons is measured and the radius calculated. The ratio, $e/m$ could be found by equating the centripetal force on the electron with force due to the movement in the magnetic field:

$$Bev = \frac{mv^2}{r}$$

But the velocity of the electrons needs to be found!

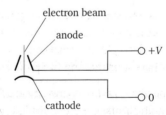

**Figure 5.78**

Figure 5.78 shows a simplified picture of the cathode–anode arrangement of the tube. Electrons produced by the cathode are accelerated towards the anode. The potential difference between the anode and cathode is $V$. Some electrons pass through the hole in the anode. These electrons retain the kinetic energy they gained by being accelerated in the electric field. The kinetic energy of the electrons is given by:

$$\tfrac{1}{2}mv^2 = eV$$

Combining this with the previous equation gives:

$$\frac{e}{m} = \frac{2V}{r^2 B^2}$$

In practice, the cathode needs to be heated to produce the electrons, and the anode is carefully staged and shaped to produce a focused beam of electrons of uniform speed. Such an arrangement is called an electron gun. A diagram of an electron gun is given in Figure 5.79. A voltage of 2 kV will produce electrons travelling at about $3 \times 10^7$ m s$^{-1}$. Details of the electron gun will not be examined.

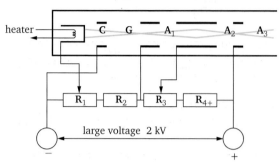

**Figure 5.79**

### Test your understanding

1 A fine beam tube accelerates electrons with a potential difference between the cathode and the anode of 250 V. The beam of electrons is 0.18 m in diameter.
  (a) Calculate the speed of the electrons.
      (Mass of electron = $9.1 \times 10^{-31}$ kg.)
  (b) Calculate the flux density of the magnetic field needed to cause the electron beam to have a diameter of 0.18 m.
      (Charge on electron = $1.6 \times 10^{-19}$ C.)

## The Mass Spectrometer

The mass spectrometer or mass spectrograph was developed in the early part of last century to determine the comparative masses of ionised atoms and, later, the relative abundance of isotopes. It uses much the same principles as those described above in relation to the cloud chamber. Figure 5.80 gives a highly simplified picture of the device.

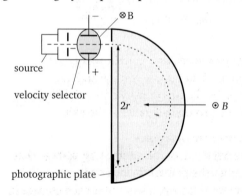

**Figure 5.80**

Positive ions produced in the chamber labelled 'source' pass into a second chamber, which is a velocity selector, that ensures that all of the ions are travelling at the same speed. In the large semi-circular chamber, the ions are subjected to another magnetic field, $B$, acting out of the plane of the diagram and at right angles to it.

$B$ forces the ions, all moving with speed $v$, into circular paths – one path is shown by the broken line on the diagram.

Once again, the force due to the charged particles' motion in the magnetic field is equated to the centripetal force on the ion. Rearranging the equation produced gives:

$$M = \frac{rqB}{v}$$

where $M$ = the mass of the ion.

All ions that have the same specific charge to mass ratio ($q/M$) travel on the same path and hit the photographic plate at the same spot. If it is known that all the ions are single ionised, for example, then we know that all ions of the same mass will hit the same spot on the photographic plate.

If, for example, the source contained singly ionised atoms of neon, containing the neon isotopes $^{20}$Ne and $^{22}$Ne, two spots would appear on the photographic plate. The spot caused by the $^{20}$Ne would be closer to the source than the spot caused by the $^{22}$Ne.

The previous example illustrates one use of the mass spectrometer. In fact, it can also indicate the relative abundance of any isotopes as the density of an image on the photographic plate is directly related to the number of nuclides striking the image spot. This use can also be applied to the quantitative analysis of mixtures of gases.

## Particle Accelerators and their Uses

The binding energies of electrons and nuclei are comparatively small – 13.6 eV (approximately $10^{-18}$ J) in the case of the hydrogen atom – and comparatively small amounts of energy, such as that available in a small direct current (d.c.), are needed to remove the electrons (ionise the atoms).

The binding energy of particles inside atomic nuclei is much greater however, and the binding of quarks inside protons and neutrons is greater still; here we are concerned with energies of the order of MeV and GeV – $10^5$ and $10^8$ times atomic binding energy.

It follows that investigating the structure of nuclei and hadrons needs something rather special. Among various devices designed and built during the first half of the twentieth century, the cyclotron and synchrotron deserve special mention on account of their usefulness and their exploitation of field theory.

## The Cyclotron

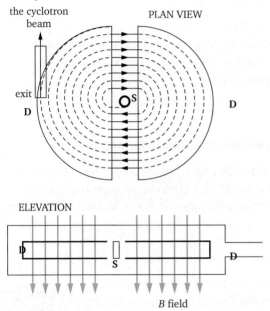

**Figure 5.81**

Figure 5.81 gives two views of the main features of a cyclotron.
- The device is enclosed in an evacuated chamber – not shown in the plan view.
- S is a source of particles – usually protons.
- D is a hollow D-shaped chamber with a magnetic field at right angles to it.
- The D-shaped chambers are connected across a high frequency, high voltage supply – not shown.

The magnetic field is used to make protons emerging at speed from the source S move in clockwise circles, when seen from above. You can check that direction using Fleming's Left Hand Rule.

The phase of the alternating supply is such that whenever the protons emerge from a particular chamber, that chamber is positive and the other chamber is negative. So each time protons cross the gap between the two chambers:
- they are accelerated...
- they gain energy equal to $qV$ at each crossing.

After each crossing:
- the speed of the protons is greater
- the radius of their circular path is greater since:

$$\frac{mv^2}{r} = Bqv$$

so:

$$\frac{r}{v} = \frac{m}{Bq}$$

In effect, the protons move in a spiral of ever-increasing radius, gaining kinetic energy at every crossing of the gap between the **Ds**.

Note that the protons move:
(i) left to right when crossing the 'upper' half of the gap
(ii) right to left when crossing the 'lower' half.

In (i), the left-hand **D** must be positive and in (ii), the right-hand **D** must be positive. In the time for each half circle, the potential difference between the **Ds** must advance a half cycle. The time, $T$, for a complete circuit must be the same as the period of the supply. If $f$ is the frequency at which the potential difference across the **Ds** changes:

$$T = \frac{1}{f} \qquad \text{but:} \quad T = \frac{2\pi r}{v}$$

substituting for $\dfrac{r}{v}$:

$$T = \frac{2\pi m}{Bq} \qquad \text{so:} \quad f = \frac{Bq}{2\pi m}$$

This is known as the *cyclotron frequency*. The maximum speed, $v_m$, of the protons is their speed in their last half circle; that is, their speed at their maximum radius, $r_m$.

From the equation at the top of the page:

$$v_m = \frac{Bqr_m}{m}$$

The maximum kinetic energy, $E_m$, is given by:

$$E_m = \tfrac{1}{2}mv_m^2 = \frac{B^2q^2r_m^2}{2m}$$

The maximum speed and kinetic energy are independent of the supply voltage and are determined by the flux density of the field and the radius of the proton half circle at the exit.

Points to note are as follows.
- It is the peak p.d. of the alternating voltage that determines the path of the protons.
- The greater this peak, the greater each change in proton speed, and the greater the increase in path radius, on each crossing of the gap.
- The greater the peak p.d., the fewer the circuits inside the **Ds**.

Wider considerations are as follows:
- Advances in science and engineering go hand in hand. Advances in one make possible further developments in the other.
- Cyclotron design and realisation are difficult and demanding, particularly with respect to precision. Only the best use of materials, measurement methods and positioning techniques make such machines possible.

At 50 MeV, an energy easily attainable in a cyclotron, the speed of a proton is approximately $10^8$ m s$^{-1}$. This is $\frac{1}{3}$ of the speed of light.

Einstein's Theory of Relativity indicates that at that kind of speed, the mass of the proton is about 5% greater than its mass when at rest. But the path radius of the proton in the cyclotron increases as its mass increases, and so does the time to complete each half circle.

At very high speeds, therefore, synchronisation of proton transits with peak voltages across the gap is not simple. Only sophisticated electronic circuitry can restore it.

The cyclotron, which was developed in the 1930s, could produce bursts of particles with energies of tens of MeV. Later developments increased these energies 1000-fold.

---

**Test your understandimg**

1   Lawrence, the inventor of the cyclotron, produced an early model with a diameter of 25 cm. It could accelerate protons up to energies of approximately 1 MeV.
Calculate:
   (a) the speed of the protons it produced
   (b) the flux density of the magnetic field that would be required
   (c) the frequency at which the p.d. across the Ds must change.
Mass of proton = $1.7 \times 10^{-27}$ kg.
Charge of proton = $1.6 \times 10^{-19}$ C.

---

## Linear Accelerators

A beam of electrons is accelerated when it passes through consecutive cylindrical electrodes that are alternately less positive and more positive, i.e. relatively negative and positive. Positively charged particles can be accelerated in just the same way if the order of the electrodes is reversed.

Large linear accelerators have been used to give particles energies as high as 50 GeV. The trick is to connect alternate electrodes to opposite sides of a high frequency, high voltage supply.

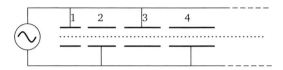

**Figure 5.82**

Figure 5.82, in illustrating the principle of the linear accelerator, shows:
- only a few of the cylindrical electrodes (1 to 4),
- the connections to the high frequency supply
- the path of the particles (central broken line).

Note that:
- the evacuated tube is not shown
- the odd-numbered electrodes are connected to one side of the supply
- the even-numbered electrodes are connected to the other side
- at peak voltages, consecutive electrodes are equally and oppositely charged
- there is no electric field within the cylindrical electrodes except near the ends
- as the particles are being accelerated, the lengths of the electrodes must increase along their path.

The linear accelerator installed at Stanford University in the USA is 3 km long and can impart energies greater than 20 GeV to electrons.

Since a 2 kV potential difference can accelerate electrons up to a speed of 10% of the speed of light, what can 20 GeV do?

$$20 \text{ GeV} = 10^7 \times 2 \text{ keV}$$

In fact, such a high potential difference can still only get electrons travelling at nearly the speed of light. This is because, at these speeds, the masses of particles increases, so they become increasingly hard to accelerate.

## The Synchrotron

The synchrotron is yet another device for accelerating charged particles to high speeds. In principle, it combines features of the cyclotron and the linear accelerator.

The particles are maintained in a circular path magnetically at constant radius; they are accelerated by means of a cylindrical anode.

Figure 5.83 shows the principle. The injection of protons, or other particles, is not shown; neither is their exit path. The path of the protons, seen to be clockwise, is indicated by the broken line.

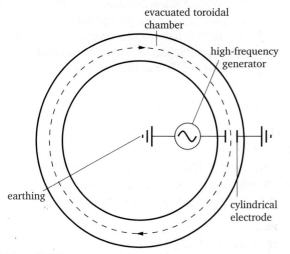

**Figure 5.83**

The large D-shaped chambers of the cyclotron are replaced by a toroidal tube (i.e. a circular tube with a circular cross-section) so smaller magnets are required.

The generator must be synchronised with the motion of the protons so that the electrode is:
- negative as the protons approach
- positive as the protons leave.

This means that the half-period of the voltage cycle must be equal to the time taken by the protons to move the length of the electrode, so...
- the frequency of the generator output increases from one proton circuit to the next.

The magnetic field strength must be synchronised with the motion of the protons so that:
- it increases as proton speed increases to maintain the same path, so...
- the magnetic field strength increases from one proton circuit to the next.

The protons make many circuits of the toroid before they are directed at their target, gaining energy on each circuit.

The principles of the above design can be improved in a number of ways, e.g. the number of electrodes may be increased.

The European Council for Nuclear Research built a synchrotron in Switzerland with a radius of 1.1 km, capable of accelerating protons to an energy of 500 GeV.

This is a developing field and energies in excess of 1000 GeV have already been obtained.

## Colliding Beams of Particles

If a particle moving with velocity $v$ strikes a stationary target, a certain amount of kinetic energy is released. If, however, it strikes a second particle which is moving with velocity $-v$, twice as much energy will be released in the collision.

So, here we have another way of increasing the energy of interactions. Once particles have been accelerated it is possible to keep them moving in circular paths and, at some convenient time introduce particles moving in the other direction.

## Investigating Particle Structures

The purpose of all these accelerators is to further our knowledge of particle properties and structures. The high energy particles we use must be able to 'see' the particles we are investigating, i.e. they must be able to interact with one another.

How do we gauge this for, a 100 GeV proton, for example? Well, we know that the particles with which we are concerned have an effective 'size' of about $10^{-14}$ m. Thinking back to our studies of Quantum Phenomena in Module 4, we recall that a moving particle possesses a wave property and that the wavelength associated with it, the de Broglie wavelength, is given by $\dfrac{h}{p}$, where $p$ is the momentum of the particle. This wavelength must be of the same order of size as the particle under investigation; just as, in diffraction, the wavelength of the diffracted wave must be a similar size to the diffracting object.

So our $\lambda$ must be about $10^{-14}$ m, and

$$\lambda = h/p$$
$$10^{-14} = \frac{6.6 \times 10^{-34}}{p}$$

or:

So, momentum $p$ is $6.6 \times 10^{-20}$ N s

The mass of a proton is $1.7 \times 10^{-27}$ kg.

So, if protons are the bombarding particles, their speed $v$ is given by:

$$v = \frac{6.6 \times 10^{-20}}{1.7 \times 10^{-27}}$$

$$v = 4 \times 10^7 \text{ m s}^{-1}$$

The kinetic energy, $\frac{1}{2} mv^2$, of such a proton is:

$$E_K = \frac{1}{2} \times 1.7 \times 10^{-27} \times 16 \times 10^{14}$$

$$= 1.4 \times 10^{-12} \text{ J}$$

$$= 8.8 \text{ MeV}$$

When using high energy particles to knock other particles to pieces – to investigate their structures – account must also be taken of the binding energies involved.

## Test your understanding

1 The diagram shows part of a mass spectrometer. A uniform magnetic field is provided by a pair of coils above and below the plane of the diagram. Helium nuclei, all of speed $v$, enter the magnetic field at **R**. They follow a semicircular path and hit the photographic plate **P**. The masses of the ions can be determined from the diameter of the path they follow.

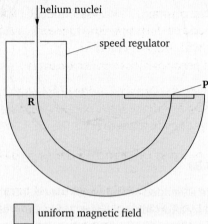

**Figure 5.84**

(a) Explain how the speed regulator works.

(b) Explain why the beam of ions moves in a circular path.

(c) State the direction of the magnetic field that would cause the movement of the helium nuclei to be as shown in the diagram.

(d) Calculate the diameter of the path followed by the nuclei.

| | |
|---|---|
| Magnetic flux density | $= 64 \text{ mT}$ |
| Speed of nuclei | $= 1.2 \times 10^6 \text{ m s}^{-1}$ |
| Charge on electron | $= -1.6 \times 10^{-19} \text{ C}$ |
| Mass of helium nucleus | $= 6.8 \times 10^{-27} \text{ kg}$ |

2 Wave particle duality suggests that a moving alpha particle (mass $6.8 \times 10^{-27}$ kg) has a wavelength associated with it. One alpha particle has an energy of $7.0 \times 10^{-13}$ J.

(a) Calculate:

(i) the momentum of the alpha particle;

(ii) the wavelength associated with the alpha particle.

(b) State and explain whether or not the alpha particle would be a suitable projectile for investigating nuclei.

# Radioactive Decay

## Radioactive Decay

The decay constant, $\lambda$, was introduced in Module 2. It is defined as the probability that any particular nucleus will decay in unit time.

$\lambda = p$ (a particular nucleus decays in unit time)

Multiplying the right-hand side of this equation by $N/N$ where $N$ is the number of nuclei of the relevant isotope present in the source:

$\lambda = \dfrac{N}{N} . p$ (a particular nucleus decays in unit time)

However, the product $N.p$ (a particular nucleus decays in unit time) is equivalent to the actual number of nuclei in the source that will decay in unit time. This is defined as the activity, $A$, of the source:

$$\lambda = \frac{A}{N}$$

or $\qquad A = \lambda N$

The units of the decay constant can be found by considering the equation:

$$A = \lambda N$$

Since $N$ is the pure number (of nuclei) and $A$ is the number (of nuclei) disintegrating per second, the units of $\lambda$ are:

$$\frac{\text{number s}^{-1}}{\text{number}}$$

units of $\lambda$: $\qquad$ s$^{-1}$

## The Radioactive Decay Curve

The equation $A = \lambda N$ fits with common sense. It says that the amount of radioactivity from a source is proportional to the amount of radioactive material that is present. You would expect this. A large block of a radioactive material is bound to be more radioactive than a piece the size of a pin head.

The equation is also useful because it leads to the familiar graph showing radioactive decay. The derivation of the equation for the graph requires calculus so you will not be expected to reproduce it. However you do need to know the resulting equations.

Since $A$ is the number of radioactive particles given off by a source in one second, it is also the same as the number of nuclei of that type that decay in one second. In other words, $A$ is equivalent to the rate of change of $N$, where $N$ is the number of that sort of nuclei present at that time:

$$A = -\frac{dN}{dt}$$

The minus sign indicates that $A$ is a positive quantity, while the rate of change of $N$ is negative (i.e. $N$ is decreasing).

Also: $\qquad A = \lambda N$

Combining these two equations:

$$\lambda N = -\frac{dN}{dt}$$

Rearranging for integration:

$$\int_{N_0}^{N} \frac{dN}{N} = \int_{0}^{t} -\lambda \, dt$$

where $N_o$ is the original number of nuclei present i.e. at a time $t = 0$.

$$\left[\ln N\right]_{N_o}^{N} = \left[-\lambda t\right]_0^t$$

Substituting the limits:

$$\ln N - \ln N_o = -\lambda t - (-)0$$

$$\ln \left(\frac{N}{N_o}\right) = -\lambda t$$

$$\frac{N}{N_o} = e^{-\lambda t}$$

$$N = N_o e^{-\lambda t}$$

This is the so-called exponential decay equation. As $t$ increases, the expression $e^{-\lambda t}$ gets smaller. The equation is saying that, at any time $t$, the number of radioactive nuclei present is equal to the original number multiplied by an expression that is getting less all the time.

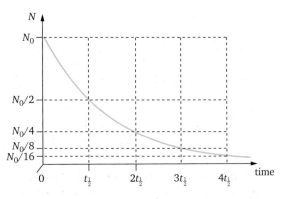

**Figure 5.85**

The most striking feature of the graph in Figure 5.85 is that each successive halving of $N$ takes the same amount of time.

This equation can be written in an alternative form:

Since:
$$N = \frac{A}{\lambda}, \quad N_o = \frac{A_o}{\lambda}$$

$$\frac{A}{\lambda} = \frac{A_o}{\lambda} e^{-\lambda t}$$

$$A = A_o e^{-\lambda t}$$

A detector measures a fixed fraction of the activity of a source. Therefore the count rate from a source is proportional to the activity. The graph of count rate against time is identical in form to the activity-time or $N$-time graph. In fact, since count rate is the parameter that is measured, in practical situations, it is most common to use that form of the graph.

### Half-life

Module 2 covered the concept of half-life, and the use of the activity against time graphs to find half-life. Now it is possible to see how half-life fits into these graphs.

The half-life, $t_{\frac{1}{2}}$, of an isotope is the time taken for the number of nuclei of that isotope to decay to half of its original number.

So:  $N = N_o e^{-\lambda t}$

but  $t = t_{\frac{1}{2}}$ when $N = \dfrac{N_o}{2}$

$$\frac{N_o}{2} = N_o e^{-\lambda t_{\frac{1}{2}}}$$

$$\frac{1}{2} = e^{-\lambda t_{\frac{1}{2}}}$$

$$2 = e^{\lambda t_{\frac{1}{2}}}$$

Taking logs of both sides of the equation:

$$\ln\ 2 = \lambda t_{\frac{1}{2}}$$

$$t_{\frac{1}{2}} = \frac{\ln\ 2}{\lambda}$$

or  $$t_{\frac{1}{2}} = \frac{0.69}{\lambda}$$

In the period $t = t_{\frac{1}{2}}$ to $t = 2t_{\frac{1}{2}}$, the number of nuclei starts as $N_o/2$ and ends as $N_o/4$. Using these data in the same decay equation:

$$\frac{N_o}{2} = \frac{N_o}{4} e^{-\lambda t_{\frac{1}{2}}}$$

$$\frac{1}{2} = e^{-\lambda t_{\frac{1}{2}}}$$

$$t_{\frac{1}{2}} = \frac{\ln\ 2}{\lambda}$$

The result is the same as for the change from $N_o$ to $N_o/2$. This shows that halving the number of nuclei takes the same length of time on each of these occasions. This works at any part of the decay, whenever the number of nuclei of that isotope halves. For example, when $N$ falls from $N_o/3$ to $N_o/6$:

$$\frac{N_o}{3} = \frac{N_o}{6} e^{-\lambda t_{\frac{1}{2}}}$$

$$\frac{1}{2} = e^{-\lambda t_{\frac{1}{2}}}$$

$$t_{\frac{1}{2}} = \frac{\ln\ 2}{\lambda}$$

This consistent result demonstrates that the half-life is a constant for any given isotope.

The relationship between half-life and decay constant also confirms the units for the decay constant.

When $t_{\frac{1}{2}}$ is given in s, the unit of $\lambda$ is $\text{s}^{-1}$.

When $t_{\frac{1}{2}}$ is given in year, the unit of $\lambda$ is $\text{year}^{-1}$.

## Using the Decay Equation

The equation can be used to find the activity at any time after a moment at which the activity was known.

EXAMPLE: $^{60}_{27}\text{Co}$ is a beta and gamma emitter with a half-life of 5.3 years. A particular $^{60}_{27}\text{Co}$ source has an activity of $3.5 \times 10^8$ Bq. What will be the activity after 16 years?

First, find the decay constant:

$$\lambda = \frac{0.69}{t_{\frac{1}{2}}}$$

$$= 0.13\ \text{year}^{-1}$$

Now simply write out the decay formula and substitute in the appropriate numbers:

$$A = A_o e^{-\lambda t}$$

$$A = 3.5 \times 10^8 \times e^{-(0.13 \times 16)}$$

$$A = 4.4 \times 10^7\ \text{Bq}$$

Alternatively, you can calculate the elapsed time for the activity to change. To do this you need to be comfortable with the use of natural logarithms. For example, how long will it take for the activity to fall to $2.5 \times 10^6$ Bq?

First find the decay constant:

$$\lambda = \frac{0.69}{t_{\frac{1}{2}}}$$

$$= 0.13 \text{ year}^{-1}$$

Next, rearrange the decay equation:

$$A = A_0 e^{-\lambda t}$$

$$\frac{A}{A_0} = e^{-\lambda t}$$

or:

$$\frac{A_0}{A} = e^{+\lambda t}$$

take natural logarithms of both sides:

$$\ln\left(\frac{A_0}{A}\right) = \lambda t$$

or:

$$t = \frac{\ln\left(\frac{A_0}{A}\right)}{\lambda}$$

$$t = \frac{\ln(3.5 \times 10^8 / 2.5 \times 10^6)}{0.13}$$

$$t = 38 \text{ year}$$

1 Cobalt-60 has a half-life of 5.3 year. When fresh, a particular source has $5.0 \times 10^{20}$ atoms of cobalt-60. It is useful until the activity has fallen below $1.5 \times 10^{12}$ Bq.
   (a) What is meant by an activity of 1 Bq?
   (b) Draw a graph showing the number of radioactive atoms over a period of three half lives.
   (c) Determine the decay constant of cobalt-60.
   (d) After what time will it be necessary to replace the source?

2 Radiocarbon dating is possible because of the presence of radioactive carbon-14 ($^{14}_{6}$C) caused by the collision of neutrons with nitrogen-14 ($^{14}_{7}$N) in the upper atmosphere. The equation for the reaction is:

$$^{14}_{7}\text{N} + ^{1}_{0}\text{n} = ^{14}_{6}\text{C} + \text{X}$$

The half-life of carbon-14 is $5.7 \times 10^3$ years.
   (a) (i) What are the proton and nucleon numbers of the particle shown as X in the above equation.
       (ii) Identify the particle X.
   (b) The mass of carbon-14 produced by this reaction in one year is 7.5 kg. 14 g of carbon-14 contains $6.0 \times 10^{23}$ atoms.
       (i) Find the number of carbon-14 atoms produced each year.
       (ii) Calculate the decay constant of carbon-14 in year$^{-1}$.
       (iii) Assuming that the number of carbon-14 atoms in the Earth and its atmosphere remains constant, then the number that decay each year is the same as the answer to (i). Use this fact and your answer to (ii) to calculate the number of carbon-14 atoms in the Earth and its atmosphere.
   (c) A sample of wood (containing carbon-14) from a tree that has recently been chopped down has an activity of 0.80 Bq. A sample of a similar size from an ancient boat had an activity of 0.30 Bq.
       (i) Draw a graph to show how the activity of carbon-14 in the sample having an initial activity of 0.80 Bq will vary over a period of three half-lives.
       (ii) Use the graph to estimate the age of the boat.
       (iii) Explain why an activity of 0.80 Bq would be hard to measure in the school laboratory.

# Appendix A

## Circular Motion and SHM

Figure A1 shows a particle **X** moving in a circular path of radius $A$.

It travels at a uniform speed $v$.

From the section on circular motion (see page 20)
- angular speed $\omega$ is $v/A$
- velocity is always tangential to the circular path so is continually changing
- period $T$ is $2\pi/\omega$.
- angular velocity $\omega$ is $2\pi f$.

The particle starts at point **P** and in a time $t$ it travels from **P** to **Q**. The angular displacement from its starting point is $\omega t$.

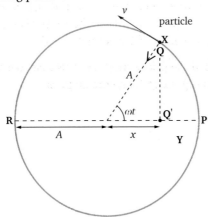

**Figure A.1**

Figure A.1 also shows a particle **Y** moving between **P** and **R** with simple harmonic motion of amplitude $A$. This particle completes one cycle in the same time $T$ as it takes **X** to complete one orbit of the circular path.

It is found that, when moving with shm, after time $t$ particle **Y** will have travelled from **P** to point **Q′** which is directly below the point **Q**. **Q′** is the projection of **Q** on to the diameter **PR**.

### Variation of displacement with time

The displacement $x$ of **Y** from its equilibrium position **O** is given by:

$$x = A \cos \omega t \qquad (1)$$

taking displacements to the right as positive.

This is the equation for shm that was quoted on page 29.

### Variation of velocity with time

The velocity of **Y** is the resolved component of the velocity of **X** along the diameter as shown in Figure A.2 (a). Taking velocities to the right as positive:

$$\text{velocity of } \mathbf{Y} = -v \sin \omega t$$
$$= A\omega \sin \omega t \qquad (2)$$

The **maximum velocity** is when $\sin \omega t$ is 1 (when $\omega t = 90°$), which is when particle **Y** passes through **O**.

The maximum velocity $= A\omega$ or $2\pi fA$, as stated on page 30.

**(a)**

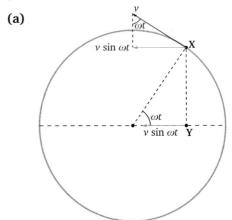

**(b)**

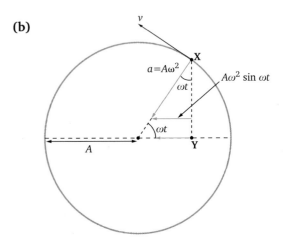

**Figure A.2**

### Variation of acceleration with time

The acceleration of **Y** is the resolved component of the acceleration of **X** along the diameter, as shown in Figure A.2 (b).

The acceleration $a$ of **X** is $\dfrac{v^2}{A}$ or $A\omega^2$ toward the centre of the circular path. Therefore:

$$a = -A\omega^2 \cos \omega t \qquad (3)$$

taking accelerations to the right as positive.

Combining equations (1) and (3) gives

$$a = -\omega^2 x$$

This is the relationship between acceleration and displacement that characterises simple harmonic motion (see page 29).

The **maximum acceleration** is when $\cos \omega t$ is 1 (when $\omega t = 0$), which is when particle **Y** is at **P** or **R**.

The maximum acceleration $= A\omega^2$ or $(2\pi f)^2 A$, as stated on page 30.

## Phase difference

In Figure A.3 particles **A** and **B** are moving in the same circular path at the same speed. **B** is following **A**.

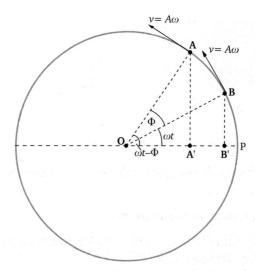

**Figure A.3**

The projections of **A** and **B** on the diameter, **A′** and **B′**, perform simple harmonic motion, with **B′** following **A′**. Particles at these points would oscillate at the same frequency and amplitude, but would be out of phase.

If timing commences when **A** is at **P** (see Figure A.3), the displacement of **A** after time $t$ is $A \cos \omega t$, the displacement of **B** is $A \cos (\omega t - \phi)$.

$\phi$ is called the **phase angle**. In this case, we say that **B** is **lagging** behind **A** by the angle $\phi$.

# Appendix B

## Spreadsheet Modelling

### Modelling principles

This approach can be applied to many time-dependent processes such as capacitor discharge, radioactive decay, oscillators and falling objects with or without a resistive force.

The design process is:
- define the initial conditions, i.e. the values of the relevant physical quantities at the start
- given these conditions, decide what physics applies to predict what happens in a given time interval $t$
- use these principles to determine the new values of the physical quantities after a short interval of time $\Delta t$
- repeat the process for the next time interval, and so on.

Two spreadsheets, for capacitor discharge and damped simple harmonic motion of a spring–mass system, have been included on the following pages to demonstrate the application of these principles. They are produced for Excel spreadsheet software and may need modification for other packages.

### Capacitor discharge

For capacitor discharge the relevant quantities are:
- the capacitance of the capacitor $C$
- the resistance of the resistor $R$
- the initial p.d. across the capacitor $V$.

### Damped SHM

In the model used, the resistive force $F_R$ is assumed to be proportional to the velocity $v$ ($F_R = av$ where $a$ is a constant that depends on the shape of the oscillating object and decides the degree of damping). This can easily be changed to assume a resistive force equal to $av^2$.

For SHM, the force toward the equilibrium position is proportional to the displacement $x$ ($F = kx$, where $k$ is the stiffness of the spring).

The relevant quantities in this case are:
- the initial maximum displacement or amplitude $A$
- the value of $a$
- the value of $k$
- the mass of the oscillating object $m$
- the initial velocity of the mass $v$

### Inserting the data

The formulae in the spreadsheet include constants and the references to the cells where the data for the calculation is to be found. The data may be one of the constants in the system being modelled, or a reference to a quantity that is assumed to remain unchanged during the particular time interval used.

Note that the shorter the time interval, the more accurately the model will match the practical results.

You must enter the cells exactly as written in the examples. When typing them out, you should complete the cells containing the initial conditions first.

### Symbols

= This must be included in order to tell the computer to calculate a value and not simply to write out text.
* This means multiply. It must always be used if the contents of two cells are to be multiplied together or if a cell's content is to be multiplied by a given number.
/ This means divide.
^ This means raise the cell contents to the power that follows (e.g. A1^2 would mean that you want to square the contents of cell A1 in the calculation).
E 1.2E3 is the same as writing $1.2 \times 10^3$.

## Some Other Useful Techniques

### Adjusting column widths

Some formulae can be very long because of the lengthy cell references.

It is useful to:
- work in font size 9 or 10. (Adjust this by highlighting the columns to be used and then adjusting the font size in the menu bar.)
- use the autoformat feature so that the columns and rows automatically adjust so that the data will fit. (Highlight cells and go to 'format' on the

menu bar; select column and autoformat). You may prefer to adjust the columns individually as you go along.

### Adjusting decimal places

This can be easily achieved by use of the $^{+.0}_{.00}$ and $^{.00}_{\rightarrow.0}$ icons on the menu bar.

### Plotting the graph

The data in the spreadsheet table can be plotted using the chart option in the spreadsheet.

To plot a graph:
- select the chart option
- select the x-y scatter option (preferably one of the 'line' options)
- select 'next'
- select 'series'
- select 'add'
- click on the icon in the right-hand side of the X values box
- use the mouse to highlight the range of X values in the table that you want to plot
- click on the icon in the right-hand side of the box when you are happy with the range
- repeat the last three steps to fix the range of Y values
- go on by selecting 'next' to add labels to the axes and grid lines to the graph or select 'finish'.

The graph can be refined by exploring the options available.

Figure B.1 shows the result of using the spreadsheet for capacitor discharge, and Figure B.2 shows the graph for simple harmonic motion using the spreadsheet models that are described on the following pages.

### Capacitor discharge

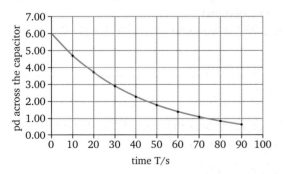

**Figure B.1**

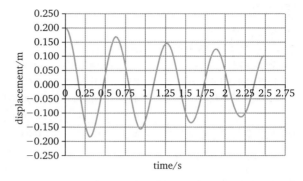

**Figure B.2**

You might begin by simply copying the spreadsheets exactly as written and then by using them to explore what happens as the initial conditions and constants for the system are changed.

You might then look closely at how the spreadsheet has been designed and try to write your own for different physical situations.

> **Tasks**
>
> 1 Modify the simple harmonic motion spreadsheet to model forced oscillations. Do this by introducing a forcing term $(p \cos 2\pi f)t$ into the force equation, which becomes:
> $$F = -kx - av + p \cos (2\pi ft)$$
> where $p$ is the peak value of the driving force in N, $f$ is the frequency at which the applied force varies.
> The values of these will need their own cell in column H.
>
> 2 Try to create spreadsheets for the following:
>   (a) an object falling freely with no air resistance
>   (b) an object falling with air resistance proportional to its velocity
>   (c) the decay of a radioactive substance.

## Typical table of results and graph

Note that the values in column G can be changed to give data for different resistors, capacitors, time intervals and initial p.d.

|  | A | B | C | D | E | F | G |
|---|---|---|---|---|---|---|---|
| 1 | p.d. across capacitor/V | discharge current/A | charge on capacitor/C | charge lost in $\Delta t$/C | time $T$/s | C/F | 0.00047 |
| 2 | 6.00 | 0.000060 | 0.00282 | 0.000600 | 0 | $R/\Omega$ | 100000 |
| 3 | 4.72 | 0.000047 | 0.00222 | 0.000472 | 10 | $\Delta t$/s | 10 |
| 4 | 3.72 | 0.000037 | 0.00175 | 0.000372 | 20 | $V_o$/V | 6.0 |
| 5 | 2.93 | 0.000029 | 0.00138 | 0.000293 | 30 |  |  |
| 6 | 2.30 | 0.000023 | 0.00108 | 0.000230 | 40 |  |  |
| 7 | 1.81 | 0.000018 | 0.00085 | 0.000181 | 50 |  |  |
| 8 | 1.43 | 0.000014 | 0.00067 | 0.000143 | 60 |  |  |
| 9 | 1.12 | 0.000011 | 0.00053 | 0.000112 | 70 |  |  |
| 10 | 0.89 | 0.000009 | 0.00042 | 0.000089 | 80 |  |  |
| 11 | 0.70 | 0.000007 | 0.00033 | 0.000070 | 90 |  |  |

The graph shown on the previous page is column **A** against column **E**.

You could also plot graphs of discharge current against time or charge remaining against time.

## Completing the spreadsheet

- You may change the column headings, but cells A2 to E2, A3 to E3 and column G should be typed in **exactly** as shown below.
- The contents of A3 to E3 may be copied into A4 to E4 and into subsequent lines as indicated in the box below.
- When a cell is identified as, for example, E2 in a formula, the cell reference will change automatically to E3 when copied into the next line.
- When a cell is included in a formula as, for example, $G$3, the actual value in this cell will be used for all calculations and will be copied to the next line as $G$3.
- You may abbreviate the headings if you wish. Do not worry if you cannot include superscripts and symbols in headings. They will not affect the data.

To copy cells:
- highlight the cells to be copied (e.g. A3–E3)
- press the right-hand mouse button
- select copy
- highlight the cells into which the copy is to be made (e.g. the block from A4–E11)
- press the right-hand mouse button
- select copy.

|  | A | B | C | D | E | F | G |
|---|---|---|---|---|---|---|---|
| 1 | p.d. across capacitor/V | discharge current/A | charge on capacitor/C | charge lost in $\Delta t$/C | time $T$/s | C/F | 0.00047 |
| 2 | =$G$4 | =A2/$G$2 | =A2*$G$1 | =B2*$G$3 | 0 | $R/\Omega$ | 100000 |
| 3 | =C3/$G$1 | =A3/$G$2 | =C2–D2 | =B3*$G$3 | =E2+$G$3 | $\Delta t$/s | 10 |
| 4 | copy of A3 | copy of B3 | copy of C3 | copy of D3 | copy of E3 | $V_o$/V | 6.0 |
| 5 | etc | etc | etc | etc | etc |  |  |

## Physical significance of the formulae and data in the cells

The spreadsheet calculates:
- the initial discharge current from $I = V/R$
- the initial charge on the capacitor from $Q = VC$
- the charge lost in time $\Delta t$ from $\Delta Q = I\,\Delta t$
- the charge left after time $\Delta t$ from $new\ Q = old\ Q - \Delta Q$
- the new p.d. across the capacitor from $V = Q/C$
- the new discharge current from $I = new\ p.d./R$
- the new charge on the capacitor from $Q = new\ p.d. \times C$
- and so on around the loop.

|  | A | B | C | D | E | F | G |
|---|---|---|---|---|---|---|---|
| 1 | p.d. across capacitor/V | discharge current/A | charge on capacitor/C | charge lost in $\Delta t$/C | time $T$/s | C/F | capacitor value |
| 2 | $V_o$ (from G4) | initial discharge current $I_o = V_o/R$ | initial charge on the capacitor $Q_o = V_o\,C$ | charge lost in first time $\Delta t$ $\Delta Q = I_o\,\Delta t$ | start time $T = 0$ | $R/\Omega$ | resistor value |
| 3 | p.d. after time $\Delta t$ $V = Q/C$ | current after time $\Delta t$ $I = V/R$ | charge after time $\Delta t$ $Q = Q_o - \Delta Q$ | charge lost in next time $\Delta t$ $\Delta Q = I\,\Delta t$ | $T + \Delta t$ | $\Delta t$/s | time interval |
| 4 | copy of A3 | copy of B3 | copy of C3 | copy of D3 | copy of E3 | $V_o$/V | initial p.d. across capacitor |
| 5 | etc | etc | etc | etc | etc |  |  |

# Simple harmonic motion

Data shows values for about $\frac{1}{4}$ of a cycle only. The graph shows the plot for many more values.

Making $c = 0$ in cell H3 will produce data for undamped simple harmonic motion.

Do not make the time interval $\Delta t$ too large as this may produce an incorrect graph showing increased amplitude with time.

To produce the modelled graph, you will need about 200 lines of data. Plot the graph using the instructions on page 137.

| | A | B | C | D | E | F | G | H |
|---|---|---|---|---|---|---|---|---|
| | time $T/s$ | displacement $x/m$ | resultant force F/N | acceleration $a/m^{-2}$ | distance moved in $\Delta t = \Delta x/m$ | velocity $v/m\,s^{-1}$ | quantity | value |
| 1 | | | | | | | | |
| 2 | 0 | 0.200 | −10.00 | −20.00 | −0.00100 | 0.000 | $k/N\,m^{-1}$ | 50 |
| 3 | 0.01 | 0.199 | −9.85 | −19.70 | −0.00299 | −0.200 | $c$ | 0.5 |
| 4 | 0.02 | 0.196 | −9.60 | −19.20 | −0.00493 | −0.397 | initial $x/m$ | 0.2 |
| 5 | 0.03 | 0.191 | −9.26 | −18.52 | −0.00682 | −0.589 | $m/kg$ | 0.5 |
| 6 | 0.04 | 0.184 | −8.83 | −17.65 | −0.00863 | −0.774 | $\Delta t$ | 0.01 |
| 7 | 0.05 | 0.176 | −8.31 | −16.61 | −0.01034 | −0.951 | | |
| 8 | 0.06 | 0.165 | −7.71 | −15.41 | −0.01194 | −1.117 | | |
| 9 | 0.07 | 0.153 | −7.03 | −14.07 | −0.01341 | −1.271 | | |
| 10 | 0.08 | 0.140 | −6.29 | −12.58 | −0.01475 | −1.412 | | |
| 11 | 0.09 | 0.125 | −5.49 | −10.98 | −0.01592 | −1.538 | | |
| 12 | 0.1 | 0.109 | −4.64 | −9.28 | −0.01694 | −1.647 | | |
| 13 | 0.11 | 0.092 | −3.75 | −7.49 | −0.01778 | −1.740 | | |
| 14 | 0.12 | 0.075 | −2.82 | −5.64 | −0.01843 | −1.815 | | |
| 15 | 0.13 | 0.056 | −1.87 | −3.74 | −0.01890 | −1.872 | | |
| 16 | 0.14 | 0.037 | −0.91 | −1.81 | −0.01918 | −1.909 | | |
| 17 | 0.15 | 0.018 | 0.06 | 0.12 | −0.01926 | −1.927 | | |
| 18 | 0.16 | −0.001 | 1.02 | 2.05 | −0.01916 | −1.926 | | |

## Completing the spreadsheet

Remember to type in formula cells exactly as shown.

| | A | B | C | D | E | F | G | H |
|---|---|---|---|---|---|---|---|---|
| | time $T/s$ | displacement $x/m$ | resultant force F/N | acceleration $a/m\,s^{-2}$ | distance moved in $\Delta t = \Delta x/m$ | velocity $v/m\,s^{-1}$ | quantity | value |
| 1 | | | | | | | | |
| 2 | 0 | = \$H\$4 | =−\$H\$2*B2−\$H\$3*F2 | =C2/\$H\$5 | =F2*\$H\$6+0.5*D2*\$H\$6^2 | =\$H\$7 | $k/N\,m^{-1}$ | 50 |
| 3 | =A2+\$H\$6 | =B2+E2 | =−\$H\$2*B3−\$H\$3*F3 | =C3/\$H\$5 | =F3*\$H\$6+0.5*D3*\$H\$6^2 | =F2+D2*\$H\$6 | $a^-$ | 0.5 |
| 4 | copy of A3 | copy of B3 | copy of C3 | copy of D3 | copy of E3 | copy of F3 | initial $x_o/m$ | 0.2 |
| 5 | etc | etc | etc | etc | etc | etc | $m/kg$ | 0.5 |
| 6 | etc | etc | etc | etc | etc | etc | $\Delta t$ | 0.01 |
| | | | | | | | $v/m\,s^{-1}$ | 0 |

## Physical significance of formulae and data in the cells

| | A | B | C | D | E | F | G | H |
|---|---|---|---|---|---|---|---|---|
| | time $T/s$ | displacement $x/m$ | resultant force F/N | acceleration $a/m\,s^{-2}$ | distance moved in $\Delta t = \Delta x/m$ | velocity $v/m\,s^{-1}$ | quantity | value |
| 1 | | | | | | | | |
| 2 | start time = 0 | $x_o$ from H4 | $F = −kx−cv$ $kx$ is the restoring force $cv$ is the damping force | $a = F/m$ | $s = ut + \frac{1}{2}a\,\Delta t^2$ $u$ in this case is the start velocity | start velocity = 0 | $k/N\,m^{-1}$ | spring stiffness |
| 3 | start time + $\Delta t$ | new displacement = old displacement + change | $F = −kx−cv$ using new F and v | $a = F/m$ using new F | $s = ut + \frac{1}{2}a\,\Delta t^2$ $u$ in this case is the velocity after $\Delta t$ | $v = u + a\,\Delta t$ $u$ is the start velocity and $v$ the final velocity | $c$ | drag factor |
| 4 | copy of A3 | copy of B3 | copy of C3 | copy of D3 | copy of E3 | copy of F3 | initial $x/m$ | amplitude |
| 5 | etc | etc | etc | etc | etc | etc | $m/kg$ | mass of object |
| 6 | etc | etc | etc | etc | etc | etc | $\Delta t$ | time interval |

# Answers

## Unit 4

### Page 10

**1** (a) $m^2\,kg\,s^{-3}\,A^{-1}$     (b) $m^{-1}\,kg\,s^{-2}$
   (c) $m^2\,kg\,s^{-3}\,A^{-2}$

**2** $kg\,s^{-2}\,A^{-1}$

**3** (a) Yes (LHS = s; RHS = s)
   (b) No (LHS = N m ; RHS = N $m^4$)

### Page 13

**1** (a) See page 20     (b) See page 21
   (c) (i) $14\,m\,s^{-1}$     (ii) $0.11\,rev\,s^{-1}$
     (iii) 39 N toward axis of rotation
   (d) 2

### Page 14

**1** $1.1 \times 10^{14}$

**2** (a) $1 \times 10^{23}$     (b) $1500\,\Omega$

**3** (a) Explain in terms of heating effect of induced
     currents.
   (b) Greater rate of change of flux in silicon, so
     greater induced current and heating effect.
   (c) High melting point since it must not melt above
     the m.p. of silicon.

**4**

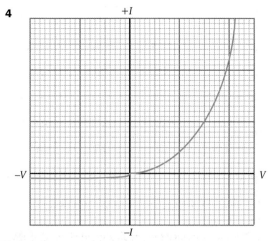

**Fig Ans 1**

**5** The applied potential difference and the thickness of
   the depletion layer.

**6** Atoms in the depletion layer are ionised. Electrons
   are available for conduction so a current flows in the
   circuit. A potential difference appears across the
   resistor. This lasts until all the ions have been
   collected producing a voltage pulse.

### Page 17

   (a) Plot ln $I$ against $t$. This should be a straight line
     of negative gradient.
   (b) $(15.5 \pm 0.2)\,m^{-1}$     (c) $(45 \pm 1)\,mm$

### Page 19

**1** (a) $k = 7.1\,kg^{\frac{1}{2}}\,s^{-1}, n = 0.5$
   (b) 19 Hz     (c) 0.083 kg

**2** $R_o = 1.4 \times 10^{-15}\,m ; n = 0.33\,(\frac{1}{3})$

### Page 22

**1** (a) 1600 N     (b) $14\,m\,s^{-1}$

**2** (a) (i) $1.6\,rev\,s^{-1}$     (ii) 0.63 s
     (iii) 2.0 m
   (c) See page 20.

**3** (a) Show direction tangential to the circular path.
   (b) (i) $1.6 \times 10^{11}\,m\,s^{-2}$     (ii) $2.7 \times 10^{-16}\,N$
     (iii)

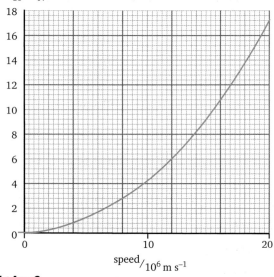

**Fig Ans 2**

**4** (a) (i) Show force $P$ acting toward the centre of
       the circle.
     (ii) Show direction **D** tangential to the circle.
   (b) (i) 1100 N     (ii) $2.7\,rad\,s^{-1}$

### Page 25

**1** (a) 600 J     (b) 460 J

**2** 640 J

**3** 54 m

**4** (**a**) 0.038 J          (**b**) 0.61 m s$^{-1}$

**5** (**a**) 0.38 J
   (**b**) Internal energy of surroundings or energy used
       to deform the rubber band permanently.

### Page 27

**1** 2.2 kW

**2** (**a**) 81 N
   (**b**) The force is balanced by air resistance.

**3** (**a**) 175 W          (**b**) 860 W
   (**c**) 98 kJ
   (**d**) (**i**) 25%          (**ii**) 53 W

**4** 94 W

**5** (**a**) 37.5 kW       (**b**) 113 kW
   (**c**) 94 kW

### Page 30

**1** 830 Hz

**2** 29 ms

**3**

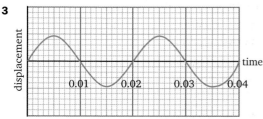

**Fig Ans 3**

**4** Amplitude = 1.5 m ; period = 0.21 s

**5** (**a**) 19.6 m s$^{-2}$       (**b**) 31 s$^{-1}$
   (**c**) 5.0 Hz

**6** (**a**) 0.47 m s$^{-1}$       (**b**) 1.48 m s$^{-2}$

### Page 33

**1** (**a**) 1.42 s          (**b**) 3.48 s

**2** (**a**) 2.12 s
   (**b**) 1.50 s (i.e., no effect on period)

**3** (**a**) 0.56 m         (**b**) 2.28 m

**4** 3.8 N m$^{-1}$

**5** 50 N m$^{-1}$

### Page 34

**2** (**a**) 1.55 m         (**b**) 32 mJ
   (**c**) 2.9 mJ

**3** 0.075 m

**4** 27 mJ

**5** (**a**) See page 29
   (**b**) Add loads to the suspended spring in small
       increments and measure corresponding
       extensions. Plot a graph of load against
       extension to determine where the graph becomes
       non-linear.

   (**c**) (**i**) 1.1 N m$^{-1}$       (**ii**) 12.6 Hz
      (**iii**) 0.59 m s$^{-1}$     (**iv**) 3.0 × 10$^{-5}$ J

### Page 39

**1** 620 N s

**2** 0.88 m s$^{-1}$

**3** (**a**) 7000 N s       (**b**) 0.58 m s$^{-1}$

**4** Increases the time taken to come to rest so that the
   force experienced by the athlete is lower compared
   with that when landing on hard ground.

**5** 12 N

**6** 0.98 N s

**7** (**a**) 1.7 N s         (**b**) 28 m s$^{-1}$
   (**c**) 120 N

### Page 41

**1** 2.5 m s$^{-1}$

**2** (**a**) 1.25 m s$^{-1}$
   (**b**) KE before = 1.8 J; KE after = 0.68 J

**3** The velocity of each is reversed. Momentum of each
   ball equal and opposite so momentum after must be
   zero. Also KE must be conserved so each ball must
   move at the same speed as before in opposite
   directions.

**4** (**a**) 60 m s$^{-1}$        (**b**) 25%

### Page 45

**1** 1.8 × 10$^5$ Pa

**2** 1.25 × 10$^5$ Pa

**3**

| | | | | |
|---|---|---|---|---|
| **b** | $p/10^5$Pa | 2.0 | 3.5 | 2.6 |
| **e** | | | | |
| **f** | $V/m^3$ | 0.15 | 0.0030 | 0.30 |
| **o** | | | | |
| **r** | | | | |
| **e** | $T/K$ | 315 | 300 | **110** |
| **a** | $p/10^5$Pa | 2.0 | **0.7** | 5.5 |
| **f** | | | | |
| **t** | $V/m^3$ | **0.062** | 0.015 | 0.40 |
| **e** | | | | |
| **r** | $T/K$ | 130 | 300 | 310 |

**4** 32 m$^3$

**5** (**a**) Take readings to show that $pV$ is constant.
   (**b**) 8.0 mol
   (**c**) For all volumes the graph should show that the
       pressure is 1.17 times greater.

**6**

| molar mass/g | 1 | 32 | 60 | 44 |
|---|---|---|---|---|
| $m$/g | 0.12 | 0.016 | 0.35 | 3.7 |
| $n$/mol | 0.12 | 5.0 × 10$^{-3}$ | 5.8 × 10$^{-3}$ | **0.083** |
| $N$ | 7.2 × 10$^{22}$ | 3.0 × 10$^{20}$ | 3.5 × 10$^{21}$ | 5.0 × 10$^{22}$ |

**7** (**a**) 3.5 mol          (**b**) 7.0 g

**8** 1200 K

**9** (a) $3.0 \times 10^{-5}$ mol    (b) $1.8 \times 10^{19}$
(c) $7.5 \times 10^{-7}$ kg

**10** (a) (i) 159°C    (ii) $1.15 \times 10^5$ Pa
(b) (i) At approximately −270°C where they intercept the temperature axis.
(ii) 0.006 mol

**11** $7.5 \times 10^{-3}$ m$^3$

## Page 50

**1** $3.1 \times 10^{-24}$ Ns

**2** $3.9 \times 10^{24}$

**3** $3.4 \times 10^{25}$

**4** (a) (i) $6.2 \times 10^{-21}$ J    (ii) $2.1 \times 10^{-20}$ J
(b) (i) 480 m s$^{-1}$    (ii) 890 m s$^{-1}$

**5** 250 m s$^{-1}$

**6** (a) $2.2 \times 10^5$ K
(b) Molecules have a range of speeds at a given temperature. Some molecules will have this speed at lower temperatures.

**7** (a) B
(b) 18.1 K (32.9 K − 14.8 K)

## Page 53

**1**

| | ΔU/kJ | Q/kJ | W/kJ |
|---|---|---|---|
| A | 200 | 100 | 100 |
| B | 0 | 50 | −50 |
| C | −50 | −300 | 250 |
| D | 120 | 520 | −400 |

**2** (a) C    (b) B and D
(c) A, B and D

## Page 55

**1** 160 J. Work is done **by** the gas.

**2** (a) 30 J of work done on the gas.
(b) 30 J of work done by the gas.
(c) 25 J of work done by the gas.
(d) 0   No change in volume so gas neither does work nor has work done on it.

**3** (a) 10.5 J    (b) energy input
(c) refrigerator

## Page 57

**1** 17 000 J

**2** 380 000 J

**3** 2800 J kg$^{-1}$ K$^{-1}$

**4** 250 s

**5** 6.1 kW

**6** (a) 7.8 K
(b) Energy lost to air as it falls: energy shared with the ground on landing.

**7** 5500 s (approximately 90 minutes)

## Page 58

**1** (a) 143 kJ    (b) 3.45 MJ
(c) 165 W

**2** (a) 1650 s
(b) Some energy from the ring is used directly to heat the air in the room. When the pan is above room temperature it will transfer energy to heat the surroundings.

**3** 350 m s$^{-1}$

**4** 0.052 kg

## Page 62

**1** (a) $1.1 \times 10^{-3}$    (b) $1.3 \times 10^5$ J m$^{-3}$

**2** 90 N

**3** (a) $9.8 \times 10^6$ Pa    (b) $4.9 \times 10^{-5}$
(c) 66 J

**4** (a) $7.2 \times 10^9$ Pa    (b) $2.3 \times 10^6$ J m$^{-3}$

**5** 1.6 J

**6** (a) $(8.5 \pm 0.5) \times 10^9$
(b) (i) $2.8 \times 10^5$ J m$^{-3}$
(ii) $8.5 \times 10^6$ J m$^{-3}$

## Page 65

**1** 2.6 V

**2** (a) 300 μF    (b) 67 μF
(c) 380 μF    (d) 22 μF

**3** (a) 220 μF    (b) 100 μF

**4** $6.0 \times 10^{-4}$ C

**5** 800 μC

## Page 66

**1**

| capacitance | p.d. | charge | energy stored |
|---|---|---|---|
| 3.2 μF | 12 V | 38 μC | 230 μJ |
| 3.4 mF | 3.5 V | 12 mC | 21 mJ |
| 4.2 μF | 24 V | 100 μC | 1.2 mJ |
| 8.0 μF | 5.0 V | 40 μC | 0.10 mJ |

**2** (a) 4.1 mA    (b) 40 mJ

**3** (a) 0.025 F    (b) 0.15 J

**4** 42 kW

## Page 68

1 Increase surface area of plates.
Decrease the separation of the plates.
Use material of higher relative permittivity between the plates.

2 (a) 150 pF  (b) 980 pF

3 (a) 2.0 mm  (b) 11 μm

4 (a) 0.36 μA
(b) Current reduced 0.09 μA

5

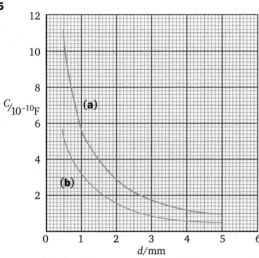

**Fig Ans 4**

## Page 72

1 (a) 5.0 V  (b) 23 μA
(c) 1.9 mC

2

| C | R | T | $T_{\frac{1}{2}}$ |
|---|---|---|---|
| 2200 μF | 100 kΩ | 220 s | 150 s |
| 470 μF | 11 kΩ | 5.0 s | 3.5 s |
| 1.3 mF | 2.2 kΩ | 2.9 s | 2.0 ms |
| 100 nF | 65 kΩ | 6.5 ms | 4.5 ms |

3 (a) 2.0 ms  (b) 90 nF
(c) 0.63 μC

4 (a) 2.2 s  (b) 1.5 s
(c) 3.0 s  (d) 2.0 s

5 (a) 6.3 V  (b) 37 μC

6 (a) 44 kΩ  (b) 0.28 mA
(c) 12 V

## Pages 76–77

1

|  | X-ray | Ultra-violet | Micro-wave | Radio-wave |
|---|---|---|---|---|
| λ/m | $1.5 \times 10^{-10}$ | $9.4 \times 10^{-8}$ | 0.040 | 200 m |
| f/Hz | $2.0 \times 10^{18}$ | $3.2 \times 10^{15}$ | $7.6 \times 10^{9}$ | $1.5 \times 10^{6}$ |
| photon energy/J | $1.3 \times 10^{-15}$ | $2.1 \times 10^{-18}$ | $5.0 \times 10^{-24}$ | $9.9 \times 10^{-28}$ |

2 If the speed falls:
(a) wavelength increases since $\lambda = \dfrac{c}{f}$.
(b) photon energy ($= hf$) is unchanged.

3 (a) Explain diffraction and interference effects.
(b) Explain evidence from photoelectric effect.

4 (a) 570 nm  (b) $8.3 \times 10^{14}$ Hz

5

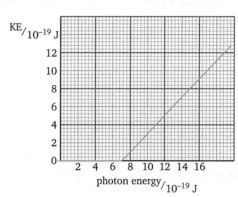

**Fig Ans 5**

6 (a) See pages 74–75  (b) 550 nm
(c) $3.6 \times 10^{-19}$ J

## Page 79

1 $4.0 \times 10^{19}$ J
2 See pages 77–78
3 See page 78
4 (a) See page 78  (b) $4.6 \times 10^{14}$ Hz
(c) 650 nm

5 (a) (i) 13.6 eV  (ii) $2.2 \times 10^{-18}$ J
(b) (i) $2.1 \times 10^{-18}$ J  (ii) $5.0 \times 10^{-20}$ J
(iii) 10

6 $1.8 \times 10^{-15}$ J (11 keV)

7 (a)

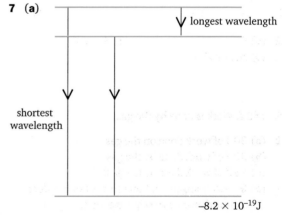

**Fig Ans 6**

(b) $4.8 \times 10^{-19}$ J; $3.9 \times 10^{-19}$ J; $0.9 \times 10^{-19}$ J
(c) $-4.3 \times 10^{-19}$ J and $-3.4 \times 10^{-19}$ J

## Page 81

1 (a) $E_1$  (b) 1.8 eV
(c) $4.4 \times 10^{14}$ Hz

2 Light from the laser is monochromatic and coherent.

3 (a) $5.1 \times 10^7$ mW m$^{-2}$  (b) $3.2 \times 10^{16}$
(c) $6.4 \times 10^5$

**Unit 5**

**Page 84**

1  (a) $4.5 \times 10^{-10}$ m  (b) $2.7 \times 10^{-23}$ N s
   (c) $1.1 \times 10^{-17}$ J

2  $3.9 \times 10^{-27}$ m s$^{-1}$

3  (a) $1.3 \times 10^{-21}$N s  (b) $7.8 \times 10^5$ m s$^{-1}$
   (c) $5.1 \times 10^{-16}$ J

4  $2.7 \times 10^{-12}$ J

5  (a) $0.5 \times 10^{-10}$ from the nucleus.
   (b) Approximately $0.25 \times 10^{-10}$m and
       $0.75 \times 10^{-10}$m from the nucleus.

6  (a) $1.4 \times 10^{-13}$ J  (b) 0.89 MeV

7  (a) $2.4 \times 10^{-10}$ m
   (b) There is no possibility of seeing any fringes.
       The angle between adjacent maximum intensity
       would be about $4 \times 10^{-5}$ radian. To produce a
       1 mm fringe spacing on a screen would require
       the screen to be 25 m away.

**Page 86**

1  (a) $5.0 \times 10^3$ V m$^{-1}$  (b) 0.033 m
   (c) $9.8 \times 10^3$ V

2  (a) $1.0 \times 10^{-15}$N  (b) $6.8 \times 10^{-17}$C
   (c) $1.4 \times 10^4$ V m$^{-1}$

3  (a) $7.0 \times 10^4$ V m$^{-1}$  (b) $2.3 \times 10^{-12}$ N
   (c) $1.4 \times 10^{-13}$ J

4  (a) $1.1 \times 10^4$ V m$^{-1}$  (b) $1.7 \times 10^{-15}$ J
   (c) $6.2 \times 10^7$ m s$^{-1}$

5  (a) F & E are vectors
   (b) F: N; PE: J; E: N C$^{-1}$ or V m$^{-1}$; V: J C$^{-1}$

**Page 89**

1  (a) $9.8$ N kg$^{-1}$  (b) 28 N
   (c) $5.2 \times 10^{-4}$ kg

2  (a) $1.9 \times 10^3$ J  (b) $1.9 \times 10^3$ J
   (c) 120 J kg$^{-1}$

3  (a) $1.7 \times 10^5$ J  (b) 63 m s$^{-1}$

4  (a) $2.5 \times 10^3$ W  (b) $2.6 \times 10^{-3}$ K

5  (a) F & g are vectors
   (b) F: N; PE: J; g: m s$^{-2}$ or N kg$^{-1}$; $V_G$: J kg$^{-1}$

**Page 91**

1  $5.8 \times 10^{-8}$ N

2  (a) $5.2 \times 10^{22}$ V m$^{-1}$  (b) $1.7 \times 10^4$ N
   (c) $5.0 \times 10^{28}$ m s$^{-2}$

3  (a) $5.3 \times 10^{-7}$ C  (b) $3.4 \times 10^{-14}$N
   (c) $2.0 \times 10^{-4}$ N

4  (a) $7.0 \times 10^{-9}$ N, $3.1 \times 10^{-9}$ N
   (b) $6.0 \times 10^{-17}$ J, $4.0 \times 10^{-17}$ J

**Page 94**

1  (a) $6.2 \times 10^3$ V m$^{-1}$ or N C$^{-1}$
   (b) $1.5 \times 10^3$ J C$^{-1}$
   (c) $1.5 \times 10^3$ V m$^{-1}$ or N C$^{-1}$
   (d) $7.7 \times 10^2$ J C$^{-1}$

2  (a) $2.6 \times 10^{-7}$ N  (b) $1.3 \times 10^{-7}$ J

3  (a) $2.0 \times 10^{-8}$ N
   (b) $5.0 \times 10^{-9}$ N, $1.25 \times 10^{-9}$ N, $3.1 \times 10^{-10}$ N
   (c) .......
   (d) 4.5 to $5.0 \times 10^{-18}$ J  (e) $4.8 \times 10^{-18}$ J

**Page 96**

**1**

| Planet | Period/$10^8$ s |
|--------|--------|
| Mercury | 0.076 |
| Venus | 0.20 |
| Earth | 0.32 |
| Mars | 0.60 |
| Jupiter | 3.7 |
| Saturn | 9.3 |
| Uranus | 26 |
| Neptune | 52 |
| Pluto | 78 |

**2**

| Planet | Speed/$10^4$ m s$^{-1}$ |
|--------|--------|
| Mercury | 4.8 |
| Venus | 3.5 |
| Earth | 3.0 |
| Mars | 2.4 |
| Jupiter | 1.3 |
| Saturn | 0.97 |
| Uranus | 0.68 |
| Neptune | 0.55 |
| Pluto | 0.48 |

**3**  Jupiter: $4.2 \times 10^{23}$ N

**4**

| Planet | Field strength/$10^{-5}$ N kg$^{-1}$ |
|--------|--------|
| Mercury | 4000 |
| Venus | 1100 |
| Earth | 600 |
| Mars | 250 |
| Jupiter | 22 |
| Saturn | 6.6 |
| Uranus | 1.6 |
| Neptune | 0.66 |
| Pluto | 0.38 |

## Page 97

**1** $3.9 \times 10^8 \, \text{m}$

**2** $2.6 \times 10^{-3} \, \text{N kg}^{-1}$

**3** $1.7 \, \text{N kg}^{-1}$

**4** $3.5 \times 10^8 \, \text{m}$

**5** (a) $4.2 \times 10^7 \, \text{m}$   (b) $3.6 \times 10^7 \, \text{m}$

**6** (a) $8.1 \times 10^6 \, \text{m}$   (b) $1.7 \times 10^6 \, \text{m}$

## Page 99

**1** (a) $-9.5 \times 10^6 \, \text{J kg}^{-1}$   (b) $-5.0 \times 10^7 \, \text{J kg}^{-1}$

**2** (a) $-2.8 \times 10^{10} \, \text{J}$   (b) $-2.7 \times 10^{10} \, \text{J}$

**3** $8.0 \times 10^9 \, \text{J}$

**4** (a) $7.3 \times 10^3 \, \text{m s}^{-1}$   (b) $8.5 \times 10^9 \, \text{J}$
   (c) $1.4 \times 10^{10} \, \text{J}$   (d) $1.5 \times 10^9 \, \text{J}$
   (e) $6.9 \times 10^9 \, \text{J}$

## Page 100

**1** (b) $5.3 \times 10^7 \, \text{J kg}^{-1}$   (c) $1.4 \times 10^{10} \, \text{J}$

**2** (c) $1.0 \, \text{N kg}^{-1}$

**3** (a) (i) 1256.2 J   (ii) 1256.8 J
   (iii) 0.05%
   (b) (i) 98 925.1 J   (ii) 98 929.3 J
   (iii) $4.2 \times 10^{-3}$ %
   (c) (i) 1 535 694.7 J   (ii) 1 535 531.3 J
   (iii) 0.01 %
   (d) (i) $4.71 \times 10^8 \, \text{J}$   (ii) $4.50 \times 10^8 \, \text{J}$
   (iii) 4.7%

**4** (a) $1.0 \times 10^{-9} \, \text{N kg}^{-1}$; $1.6 \times 10^{-10} \, \text{N kg}^{-1}$
   (b) $-4.0 \times 10^{-2} \, \text{J kg}^{-1}$; $1.6 \times 10^{-2} \, \text{J kg}^{-1}$

## Page 101–102

**1** (a) See page 85
   (b) (i) $9.5 \times 10^4 \, \text{V m}^{-1}$   (ii) $3.2 \times 10^{-19} \, \text{C}$
   (iii) Negative
   (c) $9.2 \times 10^{-20} \, \text{N}$ which is much less than
   $3.0 \times 10^{-14} \, \text{N}$

**2** (a) (i) $6.5 \times 10^{11} \, \text{V m}^{-1}$   (ii) $1.0 \times 10^{-7} \, \text{N}$
   (b) See Figure 5.19
   (c) $4.1 \times 10^{-47} \, \text{N}$ which is much less than
   $9.6 \times 10^{-8} \, \text{N}$

**3** (a) (i) $F = \dfrac{GMm}{R^2}$   (ii) See page 96
   (b) $8.1 \times 10^6 \, \text{m}$   (c) See page 97

## Page 105

**1** (a) $1.1 \times 10^{-4} \, \text{Wb}$   (b) 0.14 T
   (c) $3.3 \, \text{m}^2$

**2** $2.3 \times 10^{10} \, \text{Wb}$

**3** (a) 0.054 N   (b) $6.8 \times 10^{-3} \, \text{T}$
   (c) 0.57 m   (d) 0.22 A

**4** (a) $2.6 \times 10^{-2} \, \text{T}$   (b) $3.7 \times 10^{-5} \, \text{Wb}$

**5** $4.4 \times 10^{-3} \, \text{N}$

## Page 107

**1** (a) $9.6 \times 10^{-3} \, \text{N}$   (b) $3.8 \times 10^{-4} \, \text{Nm}$

**2** anticlockwise (viewed from above)

**3** (a) $5.8 \times 10^{-11} \, \text{N}$   (b) $5.8 \times 10^5 \, \text{m s}^{-1}$
   (c) $6.4 \times 10^{-19} \, \text{C}$   (d) 0.052 T

**4** (a) $3.5 \times 10^7 \, \text{m s}^{-1}$   (b) $3.1 \times 10^{-13} \, \text{N}$

## Page 108

**1** See page 107

**2** (a) $9.6 \times 10^{-15} \, \text{C}$   (b) 0.20 m

**3** See page 107

**4** (a) 0.77 T

**5** (a) and (b) see page 108
   (c) $7.8 \times 10^5 \, \text{m s}^{-1}$

## Page 111

**1** See page 109

**2** 0.53 V

**3** (a) $8.4 \times 10^{-3} \, \text{Wb}$   (b) $1.4 \, \text{Wb s}^{-1}$
   (c) 1.4 V

**4** 14 mV

**5** See page 110

## Page 113

**1** (a) 3.0 V   (b) 240 V
   (c) 78   (d) 104

**2** (a) 0.50 A, 6.0 W, 6.0 W
   (b) 7.2 A, 86 W, 86 W
   (c) 2.2 A, 0.037 A, 8.6 W
   (d) 0.26 A, 5.0 A, 60 W

**3** (a) 11 V   (b) 1.6 A
   (c) 17 W   (d) 0.62 (62%)

## Page 115

**1** See page 114

**2** (a) $1.1 \, \text{m s}^{-1}$   (b) $4.5 \times 10^{-2} \, \text{V}$
   (c) 1.8 V
   (d) Sine wave with max at 1.8 V and period 0.083 s.
   See page 115
   (e) Period increased by factor of 3, emf reduced to
   0.60 V. See page 115

**3** (a) $7.0 \times 10^3$   (b) See page 112

**4** (**a**) 3.5 V  (**b**) see page 112
  (**c**) see AS text: potentiometer

**5** (**a**) 220 m
  (**b**) Helium path radius 440 m in same direction as proton

## Page 117

**1** (**a**) See page 93  (**b**) See page 93
  (**c**) See page 93
  (**d**) Approx. 3300 N (from gradient of graph)

## Page 120

**1** See Figure 5.67

**2** (**a**) Very low BE – can be increased by fusion which will release energy
  (**b**) Quite high BE compared with nuclei around it – indicates stable nucleus
  (**c**) Maximum value for BE / nucleon – very stable nucleus
  (**d**) Lower than max values for BE / nucleon – can be increased by fission which will release energy

## Page 121

**1** See page 121

**2** (**a**) 10  (**b**) 60

## Page 122 (left)

**1** 0.207 u; 193 MeV

**2** (**a**) 3  (**b**) 0.266 u; 247 MeV

## Page 122 (right)

**1** (**a**) Yes: 24 MeV  (**b**) Yes: 13 MeV
  (**c**) Yes: 17 MeV

**2** $1.1 \times 10^{12}$ J

**3** See page 119

**4** Arguments related to low count rate in comparison to background.

## Page 123

**1** (**a**) (**i**) 146  (**ii**) 90
  (**b**) (**i**) $7.6 \times 10^{-30}$ kg  (**ii**) $6.9 \times 10^{-13}$ J

**2** (**a**) $1.42 \times 10^{-29}$ kg  (**b**) $1.28 \times 10^{-12}$ J

## Page 126

**1** (**a**) $9.4 \times 10^{6}$ m s$^{-1}$  (**b**) $5.9 \times 10^{-4}$ T

## Page 128

**1** (**a**) $1.4 \times 10^{7}$ m s$^{-1}$  (**b**) 1.2 T
  (**c**) 17 MHz

## Page 130

**1** (**a**) See page 108  (**b**) See page 107
  (**c**) Down into the plane of the page
  (**d**) 1.6 m

**2** (**a**) (**i**) $9.8 \times 10^{-20}$ kg m s$^{-1}$
      (**ii**) $6.8 \times 10^{-15}$ m
  (**b**) Yes: See page 129

## Page 133

**1** (**a**) See Module 2
  (**c**) 0.13 year$^{-1}$ or $4.1 \times 10^{-9}$ S$^{-1}$
  (**d**) 2.5 year

**2** (**a**) (**i**) Proton number is 1; nucleon number is 1
      (**ii**) Proton
  (**b**) (**i**) $3.2 \times 10^{26}$
      (**ii**) $1.2 \times 10^{-4}$ year$^{-1}$
      (**iii**) $2.6 \times 10^{30}$
  (**c**) (**ii**) 8000 year

# Index